Food Preservation: Principles and Practices

Food Preservation: Principles and Practices

Editor

Samiran Chavan

scitus
academics

Food Preservation: Principles and Practices

Edited by **Samiran Chavan**

ISBN: 978-1-68117-063-3

Library of Congress Control Number: 2015935461

SCITUS Academics LLC,
www.scitusacademics.com
Box No. 4766, 616 Corporate Way,
Suite 2, Valley Cottage,
NY 10989

Printed in United States of America on Acid Free Paper ∞

Contents

Preface

Food preservation usually involves preventing the growth of bacteria, fungi (such as yeasts), or other micro-organisms (although some methods work by introducing benign bacteria or fungi to the food), as well as retarding the oxidation of fats that cause rancidity. Food preservation may also include processes that inhibit visual deterioration, such as the enzymatic browning reaction in apples after they are cut during food preparation. Many processes designed to preserve food will involve a number of food preservation methods. Preserving fruit by turning it into jam, for example, involves boiling (to reduce the fruit's moisture content and to kill bacteria, etc.), sugaring (to prevent their re-growth) and sealing within an airtight jar (to prevent recontamination). Some traditional methods of preserving food have been shown to have a lower energy input and carbon footprint, when compared to modern methods. Maintaining or creating nutritional value, texture and flavor is an important aspect of food preservation, although, historically, some methods drastically altered the character of the food being preserved. In many cases these changes have come to be seen as desirable qualities – cheese, yogurt and pickled onions being common examples. The earliest form of curing was dehydration. To accelerate this process, salt is usually added. In the culinary world, it was common to choose raw salts from various sources (rock salt, sea salt, etc.). More modern "examples of salts that are used as preservatives include sodium chloride (NaCl), sodium nitrate (NaNO3) and sodium nitrite (NaNO2). Even at mild concentrations (up to 2%), sodium chloride, found in many food products, is capable of neutralizing the antimicrobial character of natural compounds.

Editor

Chapter 1

Traditional Coping Strategies and Disaster Response: Examples from the South Pacific Region

Stephanie M. Fletcher[1], Jodi Thiessen[1], Anna Gero[2], Michele Rumsey[1], Natasha Kuruppu[2], and Juliet Willetts[2]

[1]World Health Organization Collaborating Centre for Nursing, Midwifery and Health Development, University of Technology, Sydney, Broadway, NSW 2007, Australia

[2]Institute for Sustainable Futures, University of Technology, Sydney, , Broadway, NSW 2007, Australia

ABSTRACT

The Pacific Islands are vulnerable to climate change and increased risk of disasters not only because of their isolated and often low lying geographical setting but because of their economic status which renders them reliant on donor support. In a qualitative study exploring the adaptive capacity of Pacific Island Countries (PICs) across four countries, Cook Islands, Fiji, Samoa, and Vanuatu, it was clear that traditional coping strategies are consistently being applied as part of response to disasters and climate changes. This paper describes five common strategies employed in PICs as understood through this research: recognition of traditional methods; faith and religious beliefs; traditional governance and leadership; family and community involvement; and agriculture and food security. While this study does not trial the efficacy of these methods, it provides an indication of what methods are being used and therefore a starting point for further research into which of these traditional strategies are beneficial. These findings also provide important impetus for Pacific Island governments to recognize traditional approaches in their disaster preparedness and response processes.

INTRODUCTION

There exists evidence that global warming is contributing to increased numbers of observed extreme weather events, such as heat waves, extremes in precipitation, and the potential for increased number of severe storms [1]. For the Pacific, projected frequency of tropical cyclones may follow global trends of less frequent tropical cyclones and a likely increase in the relative proportion of severe storms by the end of the 21st century [2]. Coping with climate and weather extremes is a natural part of Pacific Islander's way of life. For instance, the 1997-98 El Nino events were blamed for widespread drought and famine in some Pacific islands, including major agricultural losses in Fiji, and widespread famine in parts of Papua New Guinea [3,4]. Tropical cyclones, increased rainfall, flooding, and cyclone induced storm surges have resulted in extensive infrastructural and financial losses; particularly in Samoa, Cook Islands, and Fiji [2, 5–7]. It is anticipated that with projected climate changes, patterns of extreme weather including tropical cyclones may change, with an increased proportion of severe storms [4]. The extensive coastal boundaries, the numerous islands and atolls, long coastlines, and the density of population in coastal areas of South Pacific countries have made tsunamis, floods, and tidal abnormalities the most destructive problems in the region [8, 9].

In the context of this paper, *"climate change"* refers to any change in climate over time, whether due to natural variability or as a result of human activity [10]. A UNICEF report in 2010 indicates that climate change impacts in the Asia and the Pacific region will result in mortality and injury from extreme weather events; increase in risk of water scarcity; changes in the transmission, incidence, and distribution of water-, food-, and vector-borne diseases; migration and declining livelihoods; and heightened risk of food insecurity and child malnutrition [11]. Moreover, in the Pacific Island Countries (PICs) context, there are several "nonclimate" related factors and processes which affect the ability of communities to cope with climate stresses. These include (but are not limited to) rapid population growth, land ownership and kinship issues, intrusion of western culture into traditional governance, the erosion of traditional knowledge impacting agricultural practices, natural resource degradation, and increasing reliance on imports [12, 13]. Additionally, the small population size and economic vulnerability exacerbated by frequent disasters have resulted in the need for humanitarian assistance and development aid to manage the effects of disasters and climate change across the region [14, 15]. Both Australian government, local, regional, and international nongovernmental organizations (NGOs) and multilateral organizations including United Nations agencies are key development partners and provide rapid humanitarian assistance in PICs [15–19]. Several of these organizations also work to ensure the disaster response process transitions to postdisaster development programs to enhance overall in-country capacity [6, 20–22]. Disaster preparedness, risk reduction, and climate change adaptation are some of the thematic areas of development assistance in PICs [6, 21–24].

Historically, however, Pacific populations have applied their own traditional and local knowledge to respond to environmental challenges and changes [25]. These local and traditional knowledge and skills are increasingly being recognized as vital resources for adaptation [26]. Understanding local strategies used to prepare for, respond to, and recover from climate related disasters will help to define the Pacific Islands capacity to adapt to disasters and climate related events [25, 27]. This approach has being widely recognized and adopted within the field of climate change adaptation across developing countries as a part of response [28–30]. Both international and local leaders across the Pacific recognize the importance of traditional knowledge and coping skills to combat the effects of a changing and variable climate [14, 23, 30–33]. The Pacific Islands Framework for Action on Climate Change (PIFACC) [23] recognizes and promotes the understanding and application of traditional knowledge to inform adaptation measures. To this end there has been a call for more research to link historical and anecdotal evidence of local coping strategies and for the integration of traditional knowledge into information management systems [23, 31]. Efforts have been made to capitalize on existing knowledge of regional cultural approaches and experiences with climate variability through consultative and research approaches [31]. It is therefore important for national disaster organizations and donors to recognise and incorporate traditional knowledge and skills in relevant planning and programming.

The purpose of this paper is to explore traditional coping strategies that may enhance long-term adaptive capacity to a variable and changing climate by looking specifically at examples from Pacific Islanders in their response to impacts of climate change and associated disasters. It draws on findings from a larger study which examined the adaptive capacity of the disaster response systems in four PICs to meet immediate humanitarian needs following a disaster [27]. This paper details the findings from the larger study giving examples of traditional coping strategies being used to address climate change impacts, including disasters, in terms of recognition of traditional coping strategies; faith and religious beliefs; traditional governance and leadership; family and community involvement; and agriculture and food security. These findings are useful for donor organizations and nongovernmental organizations to inform design and implementation of culturally relevant disaster risk reduction and climate change adaptation programs, in PICs and similar small island developing states.

METHODS

A qualitative research methodology was used including desktop review of the literature, policy analysis, and ninety interviews with high level professionals involved in disaster response from Australian and PIC government ministries, local and regional NGOs, and international humanitarian and donor organizations. The research process was guided by a Project Reference Group (PRG) comprising key experts from Australia and the Pacific as a form of structured stakeholder oversight and engagement in the research. Furthermore, a Conceptual Framework [27] was developed to provide the scope for this study. This framework described a cycle of adaptive learning within which the adaptive capacity of the "disaster response system" (DRS) is influenced by a range of key determinants [34]. Specific determinants of adaptive capacity were used to assess the DRS and drew upon literature spanning Earth System Governance [35], climate change adaptation [34], health resources [36–38], resilience in institutions [39], and practice theory [40]. Key determinants of adaptive capacity were identified as being interorganizational including architecture; agency; adaptiveness [35, 41]; intraorganizational objective including leadership, management, and governance structures [42]; technical capacity, tools, methods, and approaches [43]; human resource for health governance and management systems [36, 37]; and finally intra-organizational subjective including risk perceptions [34]; self-efficacy beliefs [35, 39]; elements of social practice [40]; and traditional coping strategies [40, 44]. The subjective determinant of "traditional coping strategies" is the focus of this paper and therefore expands specifically on findings regarding this key determinant [27].

A semistructured interview guide informed by a conceptual framework was used to guide face-to-face and telephone interviews with all 90 participants; and probing conducted as needed. The interview style, developed with the guidance of the PRG, was designed to appreciate cultural differences including the need for clear questions and flexibility when interviewing people from various cultural backgrounds. Relevant organizations were identified with the assistance of the PRG, with additional organizations identified through a snow ball sampling technique (Atkinson and Flint, 2001). The interviews were conducted in both Australia and four Pacific Island Countries in May–July 2012 along with workshops with in-country stakeholders to test the preliminary analysis of interviews.

Four PICs were investigated in detail to provide more of a regional viewpoint: Cook Islands, Fiji, Samoa, and Vanuatu. These countries were selected with the aim to include a variety of geographical settings (including Melanesia and Polynesia); varied policy landscapes (regarding climate change, disaster, and health-related policies and plans); mix of countries with recent significant climate-driven disasters and those in the more distant past (relating to degrees of memory and levels of preparedness); and countries that experience tropical cyclones as an example of a rapid onset disaster. The scope of the study was limited to health related humanitarian needs and as such researchers were interested in obtaining a mix of countries including high and low HRH density. Development indicators (e.g., varied Human Development Index, see UNDP, 2011) were also used as a criterion for selection to include countries at different stages of development. A full account of the methodology can be found in [27].

RESULTS AND DISCUSSIONS

Historically, the study of the traditional knowledge was the purview of anthropologists and sociologists. However, other disciplines have come to recognise the importance of the broader application and use of this knowledge before it disappears; and the value of the application of traditional methods to cope with environmental hazards is now being widely recognised [33, 45]. External influences from western colonisation, governance reforms, and development assistance have contributed to some loss of traditional and local knowledge and coping strategies [25]. Increasing commercialisation and urbanisation and the subsequent erosion of traditional social networks have also contributed to the loss of traditional knowledge and practices [13]. Within a generation traditional pass times centring on the extended family or local community have been replaced by individualised and home-based recreation [13]. Historically, livelihoods were constructed and maintained in a way that accounted for climate stress based on generations of experience of living with environmental uncertainty [12].

Members of the Pacific community recognise that with time, traditional knowledge is eroding due to a combination of factors such as migration, urbanization, and the passing away of elders [32]. It is essential therefore that beneficial traditional knowledge and cultural values are used and preserved in order to improve the ability of Pacific Islanders to build long-term resilience to natural disasters over time [32].

It became evident throughout the interviews conducted in PICs that traditional coping strategies are an important part of life and where some have been eroded, in particular with agriculture and food security, there is a current move back to integrating traditional methods, to improve disaster preparedness and response. The following section provides examples from the respondents as well as from the literature, outlining Pacific Islanders traditional coping strategies for disaster response and climate change adaptation. Table 1 summarises findings around traditional knowledge and coping strategies from the Pacific Region. However, it is recognised that continuing cultural and demographic change across the region is likely to influence traditional disaster-related coping mechanisms, including those mentioned above.

Table 1: Summary of traditional coping strategies to disasters and climate changes

Key areas	**Summary of traditional knowledge and coping strategies for response to disasters and climate change in Pacific Island Countries**
Recognition of traditional coping strategies	Secretariat of the Pacific Community's (SPC) list of crops from the climate ready collection. Glossaries of traditional climate change terms in some Pacific countries. Pacific Red Cross societies Toolkit to assess natural disasters vulnerability and response capacity.
Faith and Religious beliefs	Faith-based systems and their interventions are entrenched in the social system and can build resilience. Recognition of churches, their role in community life, including use as emergency centres or disaster shelters; provision of postdisaster counselling. Churches may be a source of resources, volunteers, welfare programs for the poor, and needy including non-members.

Traditional governance and leadership	Indication of some integration of political and traditional governance systems for disaster management. Active involvement of traditional governance structures in village life, disaster management, and climate change adaptation. Use of traditional leadership networks to share information and communicate in times of disasters. Involvement of schools, churches, and the community in disaster preparedness and response.
Family and Community Involvement	Extended family system and kinship ties provide a critical support structures in times of disaster. Movement of families from high risk areas to less vulnerable areas inland during disasters. Fostering of relationships improves coordination of response teams and helps develop good coping skills.
Agriculture and food security	Traditional agricultural disaster-preparation techniques to preserve seedlings and seeds. Documenting food preservation techniques and consideration of inclusion in school curricula. Integrated farming system approach that incorporates the planting of fruit trees with root crops. Introducing less common foods and reintroducing some forgotten foods from traditional diets. Using organic agricultural practices- pesticide free and no chemicals.

RECOGNITION OF TRADITIONAL COPING STRATEGIES

Disaster response stakeholders in four Pacific Island Countries indicated that traditional coping strategies were important to help communities cope with the effects of disasters and, increasingly, the impacts of climate change. The importance of recognising and applying traditional knowledge and coping strategies was widely recognised by interviewees. A NGO representative from Vanuatu summed it up succinctly.

"We are adaptable and will find new ways to integrate science and traditional knowledge. There are traditional skills–food security and local capacity."

Government, donor and NGO communities across all countries agreed that there was *"a lot of capacity that can be tapped into the traditional cultures ....there is great potential here because of strong communities in place."*

Unfortunately, traditional strategies were found to not always be recognised, relied on, and incorporated in national disaster response mechanisms. This may be due to the fact that "the traditional coping methods documentation is not there, it's oral." According to one regional health respondent in Vanuatu,

"There are strong partnerships between islands. This includes strong kinship ties. People develop good coping skills and foster relationships.... that are very important even during coordination of teams. We have local mechanisms in place that we (at the national level) seem to overlook. Legally, these traditional structures are not incorporated. It needs to become part of the local ways of responding."

Also a survey has indicated that the use of traditional medicine is not often acknowledged because of its association with "witchcraft" [46] (cited by WHO 2001). However, a health sector respondent indicated progress had been made with the successful publication of the traditional medicine book, Na Wai Vakaviti [47], by the Women's Association for Natural Medicinal Therapy, (Wainimate) Fiji. The respondent went on to say that *"we included the knowledge and the practice from those who have been practicing; they documented it and some those [traditional medicinal therapies] have been tested in the science lab in the University of South Pacific."*

In this study, traditional practices were chosen for illustration based on their frequency of use or discussion by responders and based on first hand experiences gained by the researchers whilst working in the region. According to WHO, there is increasing recognition amongst PICs for the need to regulate the use of traditional medicines [48, 49], although for the most part this has not yet occurred. This is due to the fact that harmful and adverse reactions can result from the use of poor quality or inappropriately administered traditional medicines and practices [49, 50]. In recognition of the importance of the use of safe and effective traditional medicinal practices and products as an important form of health care for many people in the Western Pacific Region and to maximize the potential contribution of traditional medicine to primary health care, the WHO has developed the Regional Strategy for Traditional Medicine in the Western Pacific, as a guide to relevant stakeholders [49]. However, there is the need for more empirical evidence to determine the efficacy of traditional medicine practices and products, their use in conjunction with contemporary medicine, and the need for adequate information to guide consumers to make informed decisions [49].

In recognition of the strong traditional systems in the Pacific, the Climate Change and Development (CCD) Community, of the online Pacific Solution Exchange, recently held an online discussion that called for the incorporation of traditional knowledge into national planning in order to address climate change impacts and disaster risk reduction and to balance modern technology and science with such traditional practices in the PICs. According to a member of the CCD community, "there is keen interest among both older and younger generations of islanders to record this traditional knowledge and to integrate it with scientific approaches" [32].

Advancements have already been made in the preservation of traditional knowledge. For instance, the Climate Change and Development Community reported some examples which includes list of crops from the Secretariat of the Pacific Community's (SPC) climate ready collection and glossaries of traditional climate change terms in some Pacific countries, such as the iTaukei Glossary in Fiji. There was also a call for the inclusion of traditional knowledge in school curriculums, especially in remote areas with limited or no access to the internet. There was acknowledgement by several interviewees of the need to embrace traditional knowledge as an important and essential tool for disaster risk management and to combat climate change. This will require the combination of traditional knowledge with scientific knowledge, in order to promote the environmental, economic and social sustainability necessary when working with local communities who will be the primary beneficiaries of the products. To achieve this union, careful assessment and analysis are required to ensure that innate difficulties of blending traditional knowledge with scientific knowledge are addressed and that traditional value systems are upheld [32]. For instance, the Red Cross has introduced a toolkit for Pacific Red Cross societies and island communities to assess their vulnerability and capacity in times of natural disasters. Several techniques to raise community awareness were included in the toolkit, including [51] the following:

1` recording of seasonal patterns over time,
2 documenting social changes over time that impact lives and conditions, such as local hazards, construction material and design, vegetation, livestock, and environmental changes,
3 mapping of risks, hazards, and disasters response resources,
4 sharing of knowledge between communities,
5 formal approval for traditional knowledge and disaster preparedness to be included in primary school curricula, planning, and social development programs,
6 effective risk and hazard communication between communities and local authorities.

It must be noted, however, that, in the quest to blend science and traditional knowledge, care should be taken to ensure that practices that have a deleterious effect on populations are not reinforced. This is because not all traditional coping strategies are beneficial and some, in fact, can be harmful and infringe on human rights [52]. One such example was cited by a respondent in Samoa. After the tsunami in 2009, some first responders to the affected area prioritised the removal of the dead bodies before the injured, because it was a culturally acceptable way to pay respect to the dead. Prioritising the dead over the injured, thus delaying treatment, could result in deterioration of their health status. There is limited published information about the incidence and impact of some harmful traditional and cultural practices, and in many cases, these practices may not be seen as being harmful in their culture. Engaging with local communities to determine their attitudes to certain practices as well as to undertake further studies on the incidence, causes and consequences of harmful practices are recommended.

FAITH AND RELIGIOUS BELIEFS

Pacific Islands are well known for their strong religious ties and the involvement of religious groups in the development and leadership of communities. This study found that faith (predominantly Christianity) and involvement in religious communities were an important coping strategy across the case study countries. In fact, the Council of Churches or equivalent organization in each country was an important social and cultural institution and active in disasters response. Churches were found to play an important role in community life and often used as emergency centres or shelters in times of disaster. This church-based respondent described their community disaster evacuation plan:

"Everyone is instructed that once they hear the warning they should come to the shelter (church building) with their emergency backpacks [which are already prepared with emergency supplies to last 48 hours] and walk up to the [designated] Primary School."

The role of the church is widely recognised across all levels of society. According to one NGO representative in Vanuatu, *"women's church groups are the strongest community group."* The role of the church in posttsunami response in Samoa was often highlighted by respondents. This included the role of the Church as a key agent in the provision of counselling and psychosocial support to survivors and affected families and response personnel, bringing comfort to the people and attending to the spiritual needs of community and response personnel alike. It was evident that the Church was *"held in high esteem in Samoan society."* According to one church representative,

"at the time of the tsunami in 2009, our previous chairman (deceased) was immediately contacted by the Minister for Health for comfort. The Chairman went to the main hospital and was there during the entire day comforting the health staff and the nation at large. His prayers and speeches were broadcasted live across the nation throughout the day."

International and local humanitarian organizations recognise the benefits of working with churches as a powerful social agent, to advance disaster risk reduction and climate change adaptation in PICs [53, 54]. An example of the role of the church community was given by a Samoa respondent who indicated that one church in particular was the first to be on the ground when the tsunami hit. This was possible because they had all the necessary resources in terms of relief supplies and funds. They were able to immediately mobilise land and air transportation to distribute relief items, even before the Red Cross and Government representatives came in to do their assessments. They were also reported to able to provide shelter to affected communities as well as address psychosocial needs. These comments were corroborated by other Samoan respondents.

The importance of faith to build resilience was summed up by one health respondent:

"Faith based systems and their interventions were already entrenched in the social system so people were able to turn to these during the tsunami. This is why people were more resilient than expected; they were able to recover quickly because of these social groups."

However, the role of the church in providing psychosocial and mental health support was also viewed as limited by some respondents. One regional representative in Fiji indicated that "I do not think the churches have the skills or the capacity to do that..... and the churches will only speak to the Christian population. So the other people are left out-the non-Christians." Other respondents indicated a need to have a system in place to identify persons who are isolated from these support structures such as ethnic minorities, disabled, and those who cannot go to or are not members of a local church. The extent to which aid is exclusively provided to Church members was not determined by this study, but there was no indication that this was common practice in all case study countries. Neither was it determined to what degree Church counselling was effective. However, regardless of its impact and effectiveness, Pacific Islanders see faith-based interventions as part of their lives and culturally appropriate. It is against this background that Taylor (page 177, 2003) recommended that the *"power of religious belief and of social justice in the process of healing"* is a concept that should be carefully considered by overseas responders when working in the Pacific. [55]. This study therefore recommends that further research be conducted and that opportunities for knowledge sharing be encouraged, in order to better understand the effectiveness and importance of faith-based interventions.

TRADITIONAL GOVERNANCE AND LEADERSHIP

Strong traditional local governance structures are common across Pacific Island Countries, examples of which include formal Women's Committees and Council of Chiefs [53, 56, 57]. The integration of political and traditional governance has been credited for the relative stability of governance, especially in Polynesian countries [57]. Traditional governance structures are an active part of village life and play a significant role in disaster management and climate change adaptation [56, 58]. In some cases, like Samoa, their links to the national government are formalised through hierarchical traditional governance structure [56]. Understanding and incorporating the roles and essential functions of these hierarchical structures in disaster risk reduction and climate change adaptation programs can be an essential asset [53].

In Vanuatu, for example, it was reported that communities look to the Council of Chiefs for instruction and leadership in times of disasters. A health sector respondent reported that *"they [Chiefs] use a sea shell to indicate a meeting or to prepare for a Cyclone or disaster. They also ring a bell–a different type of ringing represents different type of warnings."* Village Chiefs, who also serve as chairmen of the Area Councils (a form of local level of governance in Vanuatu), use their traditional networks in communities to share information in times of disasters. Similarly in Samoa, the Village Council is actively involved in the enforcement of adaptation activities at the community level and integration of disaster risk reduction and climate change adaptation. Evidence of this has been described through the Ministry of Women Community and Social Development's (MWCSD)Aiga ma Nuu Manuia Programme (Family Wellbeing Programme), which has gained recognition in Samoa [59]. Village Leaders reportedly ensure that *"the bell is rung to warn of impending danger such as a tsunami."* The Women's Committees at village level are also key institutions for communities both during and after disasters. A government official in Samoa reported that

"In times of disasters, The MWCSD mobilised these agents (village leaders and women's committees) at the local levels to ensure relief reaches the appropriate persons/household. They also feed into the national government emergency operations centre. The Women's groups help out at the disaster shelters cooking, feeding, counting wards."

Coordination for disaster preparedness at the village level in Samoa was reported to involve regular meetings with schools, churches, and the community to ensure that everyone is involved during disaster response. The role of traditional leadership and the active involvement of women's groups in local governance cannot be underestimated or ignored. Their involvement in the planning and implementation of disaster management and climate change adaptation programs will ensure that local knowledge and culturally appropriate approaches are applied and that there is community buy-in and support for adaptation activities.

FAMILY AND COMMUNITY INVOLVEMENT

The role of strong family ties and support systems as a strategy for coping with disasters was also identified. Respondents indicated that a strong extended family system was critical for support in times of disaster. In Fiji a government official noted that

"people are strong because of the support from the family members; so when we are here in Suva and disaster strikes somewhere else, we know exactly where our relatives are. So then we can mobilise the families here to do something and then we take it to them there. Even though the Government is going to everyone, we go and see our individual families to encourage and support them."

A NGO representative in Vanuatu summed it up as such; *"I'm surprised at the natural resilience-they[communities] can come together at the spur of the moment when they need to."*

Another example of the community coming together was described by respondents in Samoa. The Samoa Volunteer Emergency Response Team (VERT) and Water Safety Project (WSP) set up to strengthen emergency response capacity of the Samoa Fire and Emergency Services Authority (SFESA), the first responder to disasters and emergencies that requires immediate evacuation, search and rescue, first aid, and medical evacuation [60]. While in its early, capacity-building stages, the VERT-WSP programme aims to minimise dependence on permanent staff required to man base stations in the event of concurrent emergencies [60]. According to one respondent *"Post tsunami, the local farmers and carpenters [who were involved in the VERT] were some of the best actors in the response, because they were used to working in rough environments and this was an asset to search and rescue operations."*

Other social support systems include the contribution of the community to the immediate needs of the affected, a common practice in PICs. In Fiji, for example, when there is an appeal for disaster relief, people donate whatever they have "school children sometimes donate clothing and one dollar each." *This allows the government to focus on "funds for building the homes and supplying of building materials."* Another example was given from Samoa, where the fact that some families own both on the coastal lands and inland plantations, facilitating moving of people from hazard zones on the coast to safer areas inland.

AGRICULTURE AND FOOD SECURITY

Pacific communities have traditionally faced food security issues arising from extended dry periods associated with El Niño/La Niña-Southern Oscillation (ENSO) events or from the destruction of agricultural produce by cyclones and floods and over time have developed ways to preserve food to cope with such situations [61, 62] Climate change may bring about an exacerbation of such extremes in the Pacific [7]. Several respondents alluded to the fact that there was a need to return to "traditional ways of preserving food to withstand during times of crisis" with the added threats climate change may bring. However, one NGO respondent went on to say that although documentation of traditional agricultural practices is limited, "it has been passed down from generations [so] they know what to do."

There has been a call from the Climate Change and Development Community (CCDC) to renew traditional knowledge of agricultural and farming practices. In some instances, they suggested that an ideal approach could be an "integrated farming system approach" that incorporates the planting of fruit trees with root crops [32]. Examples of an integrated farming approach were cited in Samoa where it was reported that current efforts by local government organizations and NGOs are "seeking to include traditional agriculture and horticulture for food security and livelihoods through economic programs." These include how to plant, what are the best sites, organic methods, and how to cook the produce and the importance of traditional ways of growing. For example, one NGO respondent suggested that "having a backup plantation for cyclone period by growing taro patches, since bananas and breadfruit trees fall down." Existing documentation of other traditional agricultural disaster-preparation techniques come from the Pacific Climate Change and Development Community such as cutting the stems of ready to harvest cassava plants to save them from damage during a cyclone; burying the harvest of taros, so it lasts for up to six months; or preserving breadfruit and pandanus by making them into a paste in Kiribati [32]. Also, "in the old days," after a cyclone Samoan communities would salvage ready to harvest root-crop tubers, filled with valuable starch, which were then preserved in underground fermentation rooms [32].

One respondent in Fiji indicated that "knowing that climate change will bring about risks to food we have to do things to suit us and food preservation is important." To this end, it was reported that "a school teacher is preparing documents for the food security." However, the respondent went on to say that there was a need for funding to produce the documents, but there was limited funding for the food security program.

Another approach suggested by some respondents was to introduce less common foods and reintroduce some forgotten foods into diets. According to one respondent, "Cassava was traditionally perceived as pig food, but efforts are being made to re-introduce it in the Samoan diet." In Fiji, one respondent indicated that there was a risk for overfishing, and suggested alternatives. Traditionally, tuna and Skipjack are not eaten in Fiji [63], but with changing behaviour and adapting to the overfished resources, these species may become a viable alternative in Fijian diets.

Using organic agricultural practices is one way to preserve the produce. Both Fijian and Samoa respondents mentioned that the governments were supporting the growth of pesticide free and chemical free crops. One Samoan respondent indicated that while the international market wants large fruits, those that are found in the local markets are small and pesticides-free. Another regional respondent concurred; countries with strong tourism sectors such as Fiji, and increasingly Samoa and Cook Islands, have major markets for organically grown vegetables.

However, some harmful practices were identified as compounding the negative impacts of climate change. These include the use of dynamite for fishing, use of burning to clear vegetation for planting increasing the risk of soil erosion, landslides, and flooding. Similar sentiments were expressed by members of the CCDC, citing examples of maladaptation [32]. Examples of how misunderstanding of local traditional systems and practices can undermine community resilience and adaptation in times of disaster was also described. For instance, in Samoa according to one government official, after the tsunami in 2009 some NGOs were offering *"cash-for-work."* When the local authorities asked for the community's assistance later, community members requested payment. The respondent found this very unusual as previously community spirit was the default way of response. The respondent perceived this as affecting the resilience and mindset of villagers who were accustomed to working to build their own community. Furthermore, offering cash-for-work is not sustainable by the local authorities and so may have affected the adaptation capacity of the community. This is one example of how understanding the local value system is useful and how misunderstanding of such value systems may be negatively impacted by inappropriate external influences.

LIMITATIONS AND FURTHER RESEARCH

As a qualitative methodology was used, this study cannot report on the efficacy of the traditional coping strategies described. As such, these findings must be interpreted in that context. However, the rigorous methodology ensured inclusivity of a very large sample size of various sectors and stakeholders from the region, supported by literature and peer-review by subject experts from the region; providing triangulation of the broad thematic areas. Moreover the representation of Pacific Islanders on the PRG who played a critical role in informing and guiding the interpretation and analysis of research outcomes provide further validation of the findings.

Exploring a mix of countries allowed a more regional view of how disaster response plays out in different settings which proved a strength of the study. However, in terms of traditional coping strategies this broad viewpoint was also a limitation. Pacific Islands are unique in their culture and traditions, and as noted by Nunn (2009), the interpretation of the climate change concept is influenced by local culture and religion. As such traditional coping strategies that work in one country may not necessarily work the same way in another country, even across countries within the same region. As this study focussed on a regional overview, the traditional mechanisms described may be particular for the respondents' country; however, the broad thematic areas were common across all countries. Specific traditional practices for the complex and varying cultures of the Pacific would need to be explored on a very local level. Therefore further research into some of these strategies and their effectiveness in building community resilience throughout the disaster continuum is recommended.

CONCLUSIONS

This paper has provided examples of ways in which traditional practices and coping strategies for disaster response are drawn upon in the Pacific. The use of stakeholder interviews with key members of disaster response organizations in Australia and the Pacific allowed specific insight into the ways in which Pacific islanders apply traditional knowledge and coping strategies to prepare for and respond to disasters. The respondents in this study provided many examples of traditional coping strategies being used to address climate change impacts. The most important themes that emerged include recognition of traditional coping strategies; faith and religious beliefs; traditional governance and leadership; family and community involvement; and agriculture and food security. Acknowledging and preserving these useful traditional coping strategies can contribute significantly in helping the donor and NGO community in integrating them into culturally relevant programming. Furthermore they can also provide important impetus for Pacific Island governments to embrace these approaches in their disaster preparedness and response processes. While evidence of effectiveness is needed, traditional practices are still important mechanisms that are naturally integrated into the lives of Pacific Islanders as described by respondents to this study. Therefore it is important that traditional ways of operating and supporting local communities are understood by external organizations seeking to offer support and provide humanitarian response in Pacific Island Countries. Understanding the historically embedded social support systems that encourage self-help and working together is likely to enhance the outcomes of donor programs and development initiatives.

ACKNOWLEDGMENTS

This work was funded by the National Climate Change Adaptation Research Facility (NCCARF). The authors gratefully acknowledge this funding. The views expressed in this paper are those of the authors and not necessarily those of NCCARF. The authors thank all research participants for their contributions, including their Project Reference Group. This work was guided by the principles of ethics and was approved by the Human Research Ethics Committee at UTS.

REFERENCES

1. D. Coumou and S. Rahmstorf, "A decade of weather extremes," Nature Climate Change, vol. 2, no. 7, pp. 491–496, 2012.

2. T. R. Knutson, J. L. McBride, J. Chan et al., "Tropical cyclones and climate change," Nature Geoscience, vol. 3, no. 3, pp. 157–163, 2010.
3. W. Kininmonth, "The 1997-1998 El Nino event: a scientific and technical retrospective," WMO Report, 1999.
4. Australian Bureau of Meteorology and CSIRO, "Pacific Climate Change Science Program (PCCSP): current and future climate of Pacific Island countries," International Climate Change Adaptation Initiative (ICCAI), Collaboration between Pacific Osland Countries, Australian Bureau of Meteorology and Commonwealth Scientific and Industrial Research Organization, 2011.
5. A. Olsthoorn, W. Maunder, and R. S. Tol, "Tropical cyclones in the Southwest Pacific: impacts on Pacific Island countries with particular reference to Fiji," in Climate, Change and Risk, T. Downing, A. Olsthoorn, and R. Tol, Eds., pp. 221–244, Routledge, 2002.
6. J. Campbell, "Consolidating mutual assistance in disaster management within the Pacific: principles and application," in Proceedings of the 7th South Pacific Regional IDNDR Disaster Management Meeting, September, 1998.
7. K. J. E. Walsh, K. L. McInnes, and J. L. McBride, "Climate change impacts on tropical cyclones and extreme sea levels in the South Pacific—a regional assessment," Global and Planetary Change, vol. 80-81, pp. 149–164, 2012.
8. A. J. McMichael, D. H. Campbell-Lendrum, C. F. Corvalán et al., "Climate change and human health: risks and responses," Tech. Rep., WHO Library, Geneva, Switzerland, 2003.
9. ADRC, Natural Disasters Data Book 2009, 2009.
10. IPCC, "Summary for policymakers," Climate Change 2007: the Physical Science Basis, Contribution of Working Group I to the Fourth Assessment Report, Cambridge University Press, Cambridge, UK, 2007.
11. M. Urbano, N. Maclellan, T. Ruff, and G. Blashki, "Climate change and children in the Pacific Islands," UNICEF Pacific, 2010.
12. O. Warrick, "Climate change and social change: vulnerability and adaptation in rural Vanuatu," The University of Waikato, Hamilton, New Zealand, 2010.

13. M. Pelling and J. I. Uitto, "Small island developing states: natural disaster vulnerability and global change," Global Environmental Change: Part B, vol. 3, no. 2, pp. 49–62, 2001.
14. S. Fletcher, A. Gero, M. Willetts et al., "Review of Australia's overseas disaster and emergency response sector," Prepared for NCCARF, WHO Collaborating Centre and the Institute for Sustainable Futures, University of Technology, Sydney, Australia, 2012.
15. N. Maclellan, "Turning the tide: improving access to climate financing in the Pacific Islands," Lowy Institute for International Policy, 2011.
16. Australian Government, "Humanitarian preparedness and response," 2013,http://aid.dfat.gov.au/makediff/humanitarian/Pages/default.aspx.
17. Australian Government, "Regional Pacific," 2013,http://aid.dfat.gov.au/countries/pacific/rp/Pages/home.aspx.
18. CARE, "Care humanitarian action," 2011, http://www.care.org.au/Document.Doc?id=558.
19. Caritas Australia, "Annual report 2010-2011," 2011.
20. AusAID, "Pacific regional aid strategy 2004–2009," Tech. Rep., Australian Agency for International Development (AusAID), Canberra, Australia, 2004.
21. AusAID, "Integration in practice-integrating disaster risk reduction, climate change and environmental considerations in AusAID programs," Tech. Rep., Australian Agency for International Development (AusAID), Canberra, Australia, 2010.
22. Australian Government, "International climate change adaptation initiative," 2011,
23. SPREP, "Pacific Islands Framework for Action on Climate Change (PIFACC) 2006–2015," Tech. Rep., Apia, Samoa, 2007.
24. SOPAC, "Pacific disaster risk reduction and disaster management: a framework for action 2005–2015," Tech. Rep., SOPAC, Suva, Fiji Islands, 2010.
25. I. Kelman, "Hearing local voices from small island developing states for climate change," Local Environment, vol. 15, no. 7, pp. 605–619, 2010.

26. L. A. Nurse, G. Sem, J. E. Hay et al., "Small island states," Climate Change 2001: Impacts, Adaptationand Vulnerability, 2001.
27. A. Gero, S. Fletcher, M. Rumsey et al., "Disaster response and climate change in the Pacific," Tech. Rep., National Climate Change Adaptation Research Facility (NCCARF), Sydney, Australia, 2013.
28. M. K. van Aalst, T. Cannon, and I. Burton, "Community level adaptation to climate change: the potential role of participatory community risk assessment," Global Environmental Change, vol. 18, no. 1, pp. 165–179, 2008.
29. M. S. Reed, A. J. Dougill, and M. J. Taylor, "Integrating local and scientific knowledge for adaptation to land degradation: kalahari rangeland management options," Land Degradation and Development, vol. 18, no. 3, pp. 249–268, 2007.
30. A. Gero, K. Méheux, and D. Dominey-Howes, "Integrating community based disaster risk reduction and climate change adaptation: examples from the Pacific," Natural Hazards and Earth System Science, vol. 11, no. 1, pp. 101–113, 2011.
31. J. Barnett and M. Busse, "Ethnographic perspectives on resilience to climate variability in Pacific Island countries," Tech. Rep., Asia-Pacific Network (APN) for Global Change Research, Projects, 2001.
32. J. Berrel and E. Philips, "Role of traditional knowledge (TK) to address climate change impacts," Tech. Rep., Climate Change & Development (CCD) Community, of the Pacific Solution Exchange, 2013.
33. Commonwealth of Australia, "Australia's international development assistance program 2012-13," Tech. Rep., Country and Regional Programs, 2012.
34. M. Ekström, N. Kuruppu, R. L. Wilby et al., "Examination of climate risk using a modified uncertainty matrix framework—applications in the water sector," Global Environmental Change, vol. 23, no. 1, pp. 115–129, 2012.
35. F. Biermann, "'Earth system governance' as a crosscutting theme of global change research," Global Environmental Change, vol. 17, no. 3-4, pp. 326–337, 2007.

36. World Health Organization, "The world health report 2006—working together for health," Tech. Rep., World Health Organization, Geneva, Switzerland, 2006.
37. WHO/WPRO, "Human resources for health action framework for the Western Pacific Region: 2011–2015," Tech. Rep., World Health Organization Regional Office for the Western Pacific Region, Manila, Philippines, 2011.
38. K. J. Bowen, S. Friel, K. Ebi, C. D. Butler, F. Miller, and A. J. McMichael, "Governing for a healthy population: towards an understanding of how decision-making will determine our global health in a changing climate," International Journal of Environmental Research and Public Health, vol. 9, no. 1, pp. 55–72, 2011.
39. S. McManus, E. Seville, D. Brunsdon, and J. Vargo, "Resilience management: a framework for assessing and improving the resilience of organizations," Tech. Rep., Resilient Organizations Research Report, 2007.
40. Y. Strengers, "Conceptualising everyday practices: composition, reproduction and change," Tech. Rep. no. 6, Carbon Neutral Communities, 2010.
41. I. Kelman and J. J. West, "Climate change and small island developing states: a critical review,"Ecological and Environmental Anthropology, vol. 5, no. 1, pp. 1–16, 2009.
42. S. McManus, E. Seville, J. Vargo, and D. Brunsdon, "Facilitated process for improving organizational resilience," Natural Hazards Review, vol. 9, no. 2, pp. 81–90, 2008.
43. Food Agricultural Organization (FAO), Rome, Italy, 2005.
44. J. Mercer, D. Dominey-Howes, I. Kelman, and K. Lloyd, "The potential for combining indigenous and western knowledge in reducing vulnerability to environmental hazards in small island developing states," Environmental Hazards, vol. 7, no. 4, pp. 245–256, 2007.
45. J. Newton, "An assessment of coping with environmental hazards in northern Aboriginal communities," Canadian Geographer, vol. 39, no. 2, pp. 112–120, 1995.
46. K. Chen, "Fiji national workshop on traditional medicine," Mission Report 25–31, Manila, Philippines, 2001.

47. R. Singh, "Book of healing," The Fiji Times Online, 2009.
48. World Health Organization (WHO), "Legal status of traditional medicine and complementary/alternative medicine: a worldwide review," 2001.
49. WHO/WPRO, "The regional strategy for rraditional medicine in the western pacific (2011–2020)," Tech. Rep., World Health Organization Regional Office for the Western Pacific Region, Manila, Philippines, 2012.
50. WHO, "How safe is traditional medicine," 2005, http://www.who.int/features/qa/20/en/index.html.
51. N. Maclellan, "Preparing for storms: protecting the islands," Islands Business International, 2011.
52. UN ESCAP, "Harmful traditional practices in three countries of South Asia: culture, human rights and violence against women," Gender and Development Discussion Paper Series 21, United Nations Economic and Social Commission for Asia and the Pacific (UN ESCAP), Bangkok, Thailand, 2007.
53. A. Gero, K. Méheux, and D. Dominey-Howes, "Integrating disaster risk reduction and climate change adaptation in the Pacific," Climate and Development, vol. 3, no. 4, pp. 310–327, 2011.
54. Pacific Conference of Churches, "Pacific Conference of Churches Strategic Plan 2008–2012," Suva, Fiji, 2007.
55. A. J. W. Taylor, "Cross-cultural interaction in the appraisal of disaster trauma in three Pacific Island countries," Asia Pacific Viewpoint, vol. 44, no. 2, pp. 177–193, 2003.
56. E. Huffer and A. So'o, "Beyond governance in Samoa: understanding Samoan political thought,"Contemporary Pacific, vol. 17, no. 2, pp. 311–333, 2005.
57. C. Saldanha, "Strategies for good governance in the Pacific," Asian-Pacific Economic Literature, vol. 18, no. 2, pp. 30–43, 2004.
58. E. Huffer and A. So'o, "Consensus versus dissent: democracy, pluralism and governance in Samoa,"Asia Pacific Viewpoint, vol. 44, no. 3, pp. 281–304, 2003.
59. G. Kenny, "Climate resilience in Samoa: capacity assessment and enhancement consultancy," Final Report, Apia, Samoa, 2012.

60. Samoa Fire and Emergency Services Authority, "Samoa volunteer emergency response team and water safety project," Project Design Document, Apia, Samoa, 2010.
61. N. Kuruppu and D. Liverman, "Mental preparation for climate adaptation: the role of cognition and culture in enhancing adaptive capacity of water management in Kiribati," Global Environmental Change, vol. 21, no. 2, pp. 657–669, 2011.
62. J. Barr and P. Emergency, "Drought assessment: the 1997-98 El Niño drought in Papua New Guinea and the Solomon Islands," Australian Journal of Emergency Management, vol. 14, no. 2, pp. 35–41, 1999.
63. R. E. Kearney, "An assessment of the skipjack and baitfish resources of Fiji," 1984..

Chapter 2

Bacteriophages and Their Role in Food Safety

Sanna M. Sillankorva, Hugo Oliveira, and Joana Azeredo

Institute for Biotechnology and Bioengineering (IBB), Centre for Biological Engineering, University of Minho, Campus de Gualtar, 4710-057 Braga, Portugal

ABSTRACT

The interest for natural antimicrobial compounds has increased due to alterations in consumer positions towards the use of chemical preservatives in foodstuff and food processing surfaces. Bacteriophages fit in the class of natural antimicrobial and their effectiveness in controlling bacterial pathogens in agro-food industry has led to the development of different phage products already approved by USFDA and USDA. The majority of these products are to be used in farm animals or animal products such as carcasses, meats and also in agricultural and horticultural products. Treatment with specific phages in the food industry can prevent the decay of products and the spread of bacterial diseases and ultimately promote safe environments in animal and plant food production, processing, and handling. This is an overview of recent work carried out with phages as tools to promote food safety, starting with a general introduction describing the prevalence of foodborne pathogens and bacteriophages and a more detailed discussion on the use of phage therapy to prevent and treat experimentally induced infections of animals against the most common foodborne pathogens, the use of phages as biocontrol agents in foods, and also their use as biosanitizers of food contact surfaces.

INTRODUCTION

Everyday people worldwide buy and consume a diversity of products of animal and plant origin expecting these products to be safe. However, annually, millions of people become ill, are hospitalized, and die due to a variety of foodborne pathogens transmitted through foods. For instance, the World Health Organization estimated in their fact sheet of March 2007 (no. 237) that in 2005, over 1.8 million people died due to diarrhoeal diseases. Furthermore, annually in the USA alone, there are roughly 48 million illnesses, 128,000 hospitalizations, and even 3,000 deaths caused by foodborne pathogens [1].

Regardless of modern technologies, good manufacturing practices, quality control and hygiene, changes in animal husbandry, agronomic process, and in food or agricultural technology, food safety is continuously challenged by changes in lifestyle and consumer demands (e.g., ready-to-eat products) and also by the increase of international trade [2]. The most efficient means for limiting the growth of microbes are good production hygiene, a rational running of the process line, and a well-designed use of biocides and disinfectants [3]. However, even when acceptable cleaning procedures are applied, bacteria are found in foods and food contact surfaces [4]. Food products may become contaminated at different stages along the food chain, from growth or production until the final consumption. Furthermore, the inherent ability of pathogens to attach to living and inert surfaces, where they start living in microbial communities known as biofilms, and become highly tolerant to varied antimicrobial agents [5] also contributes to the pathogen prevalence in foods and food contact surfaces. So, to meet the primary goal of any food safety program, the consumer protection, new food preservation techniques have to be continually developed to meet current demands, in order to control the emerging pathogens and their impact at global scale.

BACTERIOPHAGE: CHARTING THE PATH TO FOOD SAFETY

(Bacterio)phage, viruses specifically infecting bacteria, are harmless to humans, animals, and plants. Since the discovery of phages in 1915, they have been extensively used not only in human and veterinary medicine but also in various agricultural settings. Being obligatory parasites, upon multiplication by taking over host protein machinery, phages can either cause cell lysis to release the newly formed virus particles (lytic pathway) or lead to integration of the genetic information into the bacterial chromosome without cell death (lysogenic pathway).

Towards a food safety perspective, strictly lytic phages are possibly one of the most harmless antibacterial approaches available.

Phages offer advantages as biocontrol agents for several reasons: (i) high specificity to target their host determined by bacterial cell wall receptors, leaving untouched the remaining microbiota, a property that favors phages over other antimicrobials that can cause microbiota collateral damage; (ii) self-replication and self-limiting, meaning that low or single dosages will multiply as long as there is still a host threshold present, multiplying their overall antimicrobial impact; (iii) as bacteria develop phage defense mechanisms for their survival, phages continuously adapt to these altered host systems; (iv) low inherent toxicity, since they consist mostly of nucleic acids and proteins; (v) phages are relatively cheap and easy to isolate and propagate but may become time consuming when considering the development of a highly virulent, broad-spectrum, and nontransducing phage; (vi) they can generally withstand food processing environmental stresses (including food physiochemical conditions); (vii) they have proved to have prolonged shelf life. Phages are readily abundant in foods and have been isolated from a wide variety of raw products (e.g., beef, chicken) [6, 7], processed food (e.g., pies, biscuit dough, and roast turkey) [8], fermented products (e.g., cheese, yoghurt) [9], and seafood (e.g., mussels and oysters) [8, 10]. This suggests that phages can be found in the same environments where their bacterial host(s) inhabit, or once were present and that phages are daily consumed by humans. Furthermore, the use of antibiotics prophylactically and therapeutically in farm animals has become a major concern due to their possibility of contributing to the declining efficacy of the antibiotics used to treat bacterial infections in humans and leading to the alarming emergence of superbugs likeSalmonella DT104 and the methicillin-resistant and multidrug-resistant Staphylococcus aureus. The use of phages to promote food safety can be basically done at four different stages along the food chain (Figure 1).

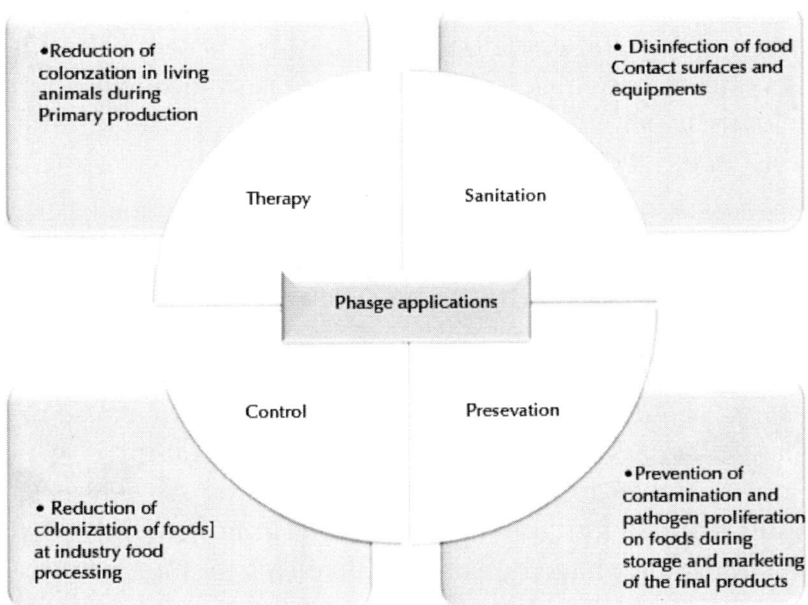

Figure 1: Feasible applications of phages along the food chain towards an increased food safety (adapted from Greer [18]).

Reduction of pathogens colonization in animals during primary production (phage therapy) is a strategy followed in primary production just before slaughter or during animal growth to reduce the probability of cross-contamination with the animal feces during food processing. For example, it is estimated that a reduction of 2 log on the Campylobacter loads in poultry intestines is sufficient to diminish 30-fold the incidence of campylobacteriosis associated with consumption of chicken meals [11]. The proof of principle of phage therapy in animals was already established for several pathogens (detailed description below). Phages can be administered orally, incorporated in drinking water or food, to control Salmonella and Campylobacterin poultry, or by spray to target avian pathogenic E. coli in poultry, and orally/rectally to control E. coli in ruminants.

Reduction of colonization on foods (biocontrol) during industrial food processing can be accomplished by applying phages directly on food surfaces, for example, in case of meats, fresh produce, and processed foods or even mixed onto raw milk. Experimental data reveals that phages are very effective against actively growing bacteria and lose effectiveness in nongrowing bacteria [12]. In these cases, effective control could be achieved, applying high titres of phages to control pathogens by "lysis from without" mechanisms [6, 13] or whenever phages start to replicate immediately after the food begins to warm (i.e., during preparation, handling, and/or consumption).

In food industry, biofilms are found on the surfaces of equipment used, for example, in food handling, storage, or processing, especially in sites that are not easy to clean or to sanitize. Some of the work using phages against in vitro biofilms formed by spoilage and pathogenic bacteria show that under ideal conditions significant viable cell reductions are achieved and thus, their use for biosanitation is promising although very challenging due to the diversity of bacteria found in different settings.

Phages are also excellent as food biopreservation agents since they are reported to lyse hosts at temperatures as low as 1°C [14, 15], limiting growth of pathogenic and spoilage bacteria on refrigerated foods (specially psychrotrophic bacteria); once the foods are taken to room temperature, phages can further control their proliferation [16].

PHAGES TARGETING DIFFERENT FOOD PATHOGENS

The global incidence of foodborne disease and costs associated are difficult to estimate, however cost at least $7 billion dollars each year in medical expenses and lost productivity in the United States according to the United States Department of Agriculture's Economic Research Service [17]. The actual figure is higher since this estimate reflects illnesses caused only by the major foodborne pathogens. The most frequently reported pathogens from animal origin, responsible for such impact, are separately discussed in this section, focusing on their contamination sources and their harms for animals and humans and furthermore summarizing the recent phage interventions and main outcomes. We also briefly address some of the common phytopathogens and discuss phage applications carried out in an attempt to decrease crop diseases and product loss.

FOODBORNE PATHOGENS FROM ANIMAL ORIGIN

The four main foodborne pathogens from animal origin are accounted to be E. coli, Campylobacter, Salmonella, and Listeria. These bacteria are all common contaminants of ruminants, poultry, and swine and are usually carried in their gastrointestinal tract asymptomatically. Research on the use of phages against foodborne pathogens from animal origin has mainly focused on the optimization of preharvest interventions where the phage administration routes and delivery processes have received most attention and also on the optimization of postharvest strategies. The usage of phages as a preharvest strategy is made directly by administering phages to livestock to prevent animal illness and/or also to minimize the pathogen carriage in the gastro-intestinal tract, thereby preventing pathogen entry to the food supply. Postharvest strategies are based on the use of phages directly on animal carcasses in an attempt to sanitize the products.

Escherichia coli is a gram-negative bacterium. Serotype O157: H7 in particular, classified as Shiga toxin-producing E. coli, is a well-known food poisoning pathogen. Its major reservoir comprises ruminants and, as it can survive well under intestinal conditions, if proper care is not taken during slaughter, the contents of the intestines, fecal material, or dust on the hide may contaminate meats [19]. The most common route of E. colitransmission to humans is via undercooked contaminated food, while water and raw milk are assumed to be related to cross-contamination events, by direct or indirect contact with feces. This microorganism is highly virulent and a public health threat because ingestion of a concentration as low as 10 cells is able to cause infection [1, 20, 21].

Recent pre- and postharvest phage research targeting E. coli is listed in Table 1.

Table 1: Pre- and postharvestE. coliO157 : H7 phage applications

Year	Animal/product	Phage(s)	Strategy	Main outcome	Refs
Preharvest application					
2002	Poultry (broiler chicken)	SPR02	Air sac or drinking water	Air sac inoculation prevented mortality. Drinking water offered no protection	[24]
2002	Poultry (broiler chicken)	SPR02 and DAF6	Sprayed	Significant decrease of mortality but not complete protection	[22]
2003	Poultry (broiler chicken)	SPR02 and DAF6	Sprayed and i.m. injection	Aerosol spray effective only when applied immediately after bird challenge with E. coli. A single i.m. injection reduced mortality when applied immediately and 24 and 48 h after challenge	[23]
2003	Ruminant (lamb)	DC22	Oral delivery	No reduction of fecal shedding over 30 days	[29]
2006	Poultry (broiler chicken)	SPR02 and DAF6	i.m. injection into the left thigh	Only high phage titers (10^8) reduced mortality	[30]
2006	Ruminant (sheep)	CEV1	Oral delivery	2 log CFU reduction within 2 days	[31]
2006	Ruminant (cattle)	Phage cocktail (KH1, SH1)	Oral/rectal delivery (via drinking water)	No reduction of CFU when applied orally. Combined oral/rectal treatment reduced CFU but did not eradicate it	[32]
2009	Ruminant (steer)	Phage cocktail	Oral/rectal delivery	Small fecal shedding reduction of oral/rectal compared to the rectal treatment and control	[33]

2010	Ruminant (steer)	Phage cocktail (wV8, rV5, wV7, wV11)	Oral delivery (gelatin capsules and in feed)	No reduction of fecal shedding of nalidixic acid-resistant E. coliO157 : H7, but duration of shedding was reduced by 14 days	[28]
2010	Poultry	Phage cocktail	Oral delivery and spray	Significant reduction of mortality in large scale animal experiments	[34]
2010	Ruminant (cattle)	Phage cocktail (e11/2, e4/1c)	Oral delivery	Rapid CFU decrease within 24 to 48 h, but no decrease in fecal shedding levels	[35]
2011	Ruminant (sheep)	Phage cocktail (CEV1, CEV2)	Oral delivery	Cocktail eradicated (>99.9%) the pathogen and is more effective than CEV1 alone	[36]
Postharvest application					
2004	Meat	e11/2, e4/1c, pp01	Applied on top	Eradication in seven of nine samples	[37]
2008	Fresh produce (tomato, spinach) and meat	Phage cocktail (ECP-100)	Applied on top/sprayed	94% and 100% reductions in CFU after 120 h and 24 h in tomato and spinach; 95% reduction in ground meat after 24 h at 10°C	[38]
2009	Fresh produce (lettuce, cantaloupe)	Phage cocktail (ECP-100)	Sprayed	Significant CFU reductions after 2 days at 4°C	[39]
2011	Fresh produce (lettuce, spinach)	Phage cocktail (BEC8)	Added to foods together with trans-cinnamaldehyde (TC)	No survivors detectable after 10 min of phage combined with the TC treatment	[40]
2011	Food surfaces (spinach blades)	Phage cocktail	Sprayed	4.5 log reduction CFU after 2 h of phage	[41]
2011	Food surfaces (steel, ceramic chips)	Phage cocktail (BEC8)	Applied on top	Eradication after 10 min at 37°C and after 1 h at 23°C	[40]

Recent phage therapy to decrease E. coli levels on farm animals has focused mainly on poultry and ruminants.

Application of phages to poultry has been successful to prevent fatal respiratory infections in broiler chickens [22–24]. Several different approaches have been used; however, aerosol spraying and intramuscular (i.m.) injection have given the best results and reduced significantly the mortality of broiler chicken. Despite these results, phage administration via addition to bird drinking water proved to be inefficient in protecting the birds from fatal E. coli respiratory infections.

Although some successful results of phage therapy in ruminants have been reported using oral delivery of phages via direct administration or addition to drinking water and/or feed, the majority of the recently published papers suggest that oral treatment is unsuccessful in reducing E. coli levels (see Table 1).

The main speculated causes for the failure of oral treatments have been reported to be (i) nonspecific binding of phages to food particles and other debris in the rumen and gastrointestinal tract [25]; (ii) phage inactivation upon contact with the acidic conditions of the abomasum [26]; (iii) causing an insufficient number of orally phages reaching the gastrointestinal tract [27]. An interesting approach to reduce coliphage inactivation has been described by Stanford and colleagues in 2010. These authors successfully encapsulated phages in polymeric matrices which resisted in vitro acidic conditions and furthermore, once delivered orally to steers caused reduction of E. coli levels [28].

Essentially, oral delivery of phages has reported to be successful either using a cocktail of phages or combined with rectal treatment. For instance, a combined oral/rectal treatment using phages KH1, SH1 was able to reduce E. coli levels compared to oral treatment alone or to individual phage administration, but still this combined oral/rectal treatment using a phage cocktail did not cause total eradication of the pathogen from cattle [32]. Also, a cocktail of phages CEV1 and CEV2 were reported to lead to more than 99.9% of reduction of E. coli in sheep guts [36]. However, although reductions of E. coli levels in different organs have been described, phages still fail in reducing fecal shedding [29, 35], and only in one published work the authors managed to reduce fecal shedding duration by 14 days [28]. In 2008, Niu and colleagues investigated two different sampling techniques: fecal grab and rectoanal mucosal swab for surveillance of E. coli O157 : H7 [42] and to study the role of phage as a mitigation strategy. Their study showed discrepancy between fecal and rectoanal mucosal swab sampling in 63 of the 213 positive samples from experimentally inoculated feedlot steer. This shows that sampling procedures can significantly influence the merit of phage and that prudence must be taken to validate phage therapy for E. coli O157 : H7 control [42].

All postharvest interventions reported since 2000 have been effective in reducing E. coli levels from fresh produce and meats. Also, phage application to E. coli contaminated food contact surfaces has led to significant reductions proving that phage can be used safely for equipment and food contact surface sanitation.

The successful use of coliphages has led to the development of a phage-product (EcoShield (Intralytix)) which received regulatory approval from the Food and Drug Administration (FDA) in 2011. This product can be used on red meat parts and trims intended to be ground (Food Contact Notification no. 1018) and has proved to eliminate from 95 to 100% of E. coli O157: H7. Finalyse (Elanco Food Solutions) is another product on the market using naturally occurring phages specific for E. coli O157 : H7. Finalyse is sprayed on cattle to reduce the load of E. coli, prior to its entering the beef packing facility (preharvest strategy).

Campylobacter is a genus of gram-negative, spiral, motile, and microaerophilic bacteria with an optimal growth temperature around 41°C. C. jejuni and C. coli members are considered to be major aetiological agents of enteric diseases worldwide. Campylobacter is the most commonly reported zoonosis in Europe (EFSA 2011), and C. jejuni, in particular, is estimated to cause approximately 845,000 illnesses, 8,400 hospitalizations, and 76 deaths each year in the USA [1]. This widespread infection is explained because ingestion of low doses (400–500 cells) [43] can cause campylobacteriosis typically characterized by fever, bloody diarrhea, and acute abdominal pain [44]. Campylobacter is capable of colonizing the intestine of poultry and cattle, and thus infection is mostly acquired by fecal-oral contact, ingestion of contaminated foods (i.e., raw meat and milk contaminated through feces), and waterborne through contaminated drinking water [45–47]. The widespread disease and economic impact on agriculture and food industries has led to the development of various approaches to contain this infection using bacteriophages (Table 2).

Table 2: Pre- and postharvest Campylobacter phage applications

Year	Animal/product	Phage(s)	Strategy	Main outcome	Refs
Preharvest application					
2005	Poultry (broiler chickens)	ΦCP8, ΦCP34	Oral delivery in antacid suspension	Decrease of CFU between 0.5 and 5 log CFU/g in the cecal content over a 5-day period posttreatment	[48]
2005	Poultry (broiler chickens)	69, 71	Oral delivery	Reduction of CFU by 1 log within 5-day period post-treatment. Phage preventive treatment caused a delay in a colonization	[58]
2009	Poultry (broiler chickens)	CP220	Oral delivery	Reduction of 2 log CFU per g in cecal content after 48 h inoculated C. jejuniand C. coli birds with a single dose (7 log PFU) of CP220	[49]
2010	Poultry (broiler chickens)	Phage cocktail	Oral delivery (oral gavage and in feed)	Reduction levels of C. coli and C. jejuni in feces by 2 log CFU per g when administered by oral gavage and in feed	[50]
Postharvest application					
2003	Meat (chicken skin)	Φ2	Applied on top	1 and 2 log CFU reduction at 4°C using 10^7 PFU per mL. 10^5 and 10^3 PFU per mL failed to decrease CFU	[58]
2003	Meat (chicken skin)	Φ29C	Applied on top	MOI 1 caused less than 1 log reduction in CFU; MOI 100–1,000 caused 2 log reductions in CFU	[49]

2008	Meat (raw and cooked beef)	Cj6	Applied on top	The largest reductions were recorded at high host cell density on both raw and cooked beef over a period of 8 days incubation at 51°C	[16]

Phage preharvest interventions reported so far have been successful in reducing Campylobacter numbers in the cecal content and feces of experimentally infected broilers and have not caused any adverse health effects. However, some degree of resistant phenotypes, recovered from phage-treated chickens, has been reported [48–50], of which some were found to display clear evidence of genomic rearrangements [51]. Scott and colleagues demonstrated that Campylobacter virulent phages have the potential to activate dormant prophages, leading to rapid pathogen evolution towards the development of phage-resistant phenotypes. These authors also showed that whilst pathogen evolution can be rapid, resistance to the therapeutic bacteriophage is associated with a decreased fitness to environment, specifically meaning that phage-resistant phenotypes exhibited a decreased ability to colonize the gastrointestinal tract [51]. These genomic instability findings suggest that C. jejuni adopt this strategy to temporarily survive local environmental pressures, including phage predation and competition for resources. The key elements for the success of phage therapy against Campylobacter in broiler chickens are a proper selection of phage, the dose of phage applied, and the time elapsed after administration [48].

As for postharvest strategies, two studies have been reported (see Table 2) in which phages reducedCampylobacter contamination following a "lysis from without" mechanism. This suggests that Campylobacterphages can be used as a tool for biocontrol purposes.

It is known that Campylobacter attaches and forms biofilms on surfaces as a measure to overcome environmental stresses, such as aerobic conditions, desiccation, heating, disinfectants, and acidic conditions frequently encountered in food environments [52]; however, to date, we were only able to find one report evaluating the efficacy of Campylobacter phages of disrupting biofilm formed on glass. In this study, phages were able to reduce by 1 to 3 log the viable cell counts under microaerobic conditions; however, after treatment above 84% of the surviving bacteria were resistant to the two phages applied [53].

Salmonella, is a genus of gram-negative facultative intracellular species, is considered to be one of the principal causes of zoonotic diseases reported worldwide. Salmonella serovars can colonize and persist within the gastrointestinal tract, and so human salmonellosis is commonly associated with consumption of contaminated foods of animal origin. Salmonella infections cost nearly 3 million euros in

EU per annum in health care systems [17, 54]. Salmonella enterica serovars, Enteritidis and Typhimurium, are responsible for the majority of Salmonella outbreaks, and most events are reported to be due to consumption of contaminated eggs and poultry, pig, and bovine meats, respectively [55]. Salmonella is also a known spoilage bacterium in processed foods. Once ingested, this microorganism can cause fever, diarrhea, abdominal cramps, and even life-threatening infections [56]. To prevent such infections, a number of studies on animal phage therapy have been reported where phages were used to prevent or reduce colonization and diseases in livestock. All recent studies are summarized in Table 3.

Table 3: Pre- and postharvest Salmonella phage applications

Year	Animal/product	Phage(s)	Strategy	Main outcome	Refs
Preharvest application					
2001	Poultry (chicken)	Phage cocktail	Oral delivery (direct and via feed)	Reduction of CFU in cecal counts between 0.3 and 1.3 log compared to controls birds	[65]
2001	Swine (pig)	Felix01	Oral delivery and i.m.	Reduction of CFU in the tonsils and cecum	[66]
2005	Poultry (broiler chickens)	CNPSA1, CNPSA3, CNPSA4	Oral delivery	Reduction of CFU by 3.5 orders of magnitude after five days	[67]
2005	Poultry (chickens)	Phage cocktail (Sa2, S9, S11)	Oral delivery phage/competitive exclusion	Reduction of CFU in cecum and ileum after phage cocktail and/or competitive exclusion treatment	[68]
2007	Poultry (broiler chickens)	Φ151, Φ25, Φ10	Oral delivery (antacid suspension)	Reduction of 4.2 log and 2.19 log with phages Φ151 and Φ25 within 24 h compared with control	[69]
2007	Poultry (broiler chickens)	Phage cocktail (CB4ϕ, WT45ϕ)	Oral delivery	Reduction of CFU in cecal tonsils after 24 h. No significant differences at 48 h compered to controls	[70]
2008	Poultry (chickens)	Phage cocktail	Oral delivery (coarse spray/drinking water)	Reduction of intestinal colonization of ten-day-old experimentally contaminated birds	[71]
2010	Swine (pig)	Phage cocktail		Reduction of colonization by 99.0 to 99.9% in the tonsils, ileum, and cecum	[72]

2011	Swine (weaned pigs)	Phage cocktail	Oral delivery	Significant reduction of CFU in the rectum	[73]
2011	Poultry (chickens)		Oral delivery (via feed)	Phage prevented horizontal transmission on six-week-old infected chickens	[74]
Postharvest application					
2001	Processed food (ripened cheese)	SJ2	Added to milk	No survival during 89 days in pasteurized cheeses containing phages (MOI 10^4)	[75]
2001	Fresh produce (fresh-cut melon and apple)	Phage cocktail	Added to foods	Significant CFU reduction on melon but not on apple	[76]
2003	Meat (chicken skin)	P22, 29C	Applied on top	MOI 1 caused less than 1 log reduction in CFU; MOI 100-1,000 caused 2 log reductions in CFU and eradicated resistant strains	[13]
2003	Meat (chicken frankfurters)	Felix O1		Approx. 2 log reduction with a MOI of 1.9×10^4	[21]
2004	Fresh produce (sprouting seeds)	A, B	Applied by immersion	Phage-A reduced CFU by 1.37 logs on mustard seeds. Cocktail resulted in a 1.5-log reduction in CFU in the soaking water of broccoli seeds	[77]
2005	Meat (broiler, turkey)	PHL 4	Sprayed	Phage treatments reduced frequency of Salmonella recovery as compared with controls	[78]
2008	Meat (raw/cooked beef)	P7	Applied on top	Reduction in CFU of 2-3 log at 5°C and approx. 6 log at 24°C	[16]

2009	Fresh produce (tomatoes)	Phage cocktail	Phage + E. asburiae JX1 added to food	Prevalence reduction of internalized S. Javiana, although the major suppressing effect was via antagonistic activity of E. asburiae JX1	[79]
2010	Fresh produce (mung bean sprouts and alfalfa seeds)	Phage cocktail	Phage + E. asburiae JX1 added to foods	Combined biocontrol with E. asburiae and phage suppressed pathogen growth on mung beans and alfalfa seeds	[80]
2011	Meat (pig skin)	Phage cocktail (PC1)	Applied on top	Above 99% reduction in CFU for MOI of 10 or above at 4°C for 96 h	[59]
2012	Ready-to-eat foods and chocolate milk	FO1-E2	Added to foods and mixed in milk	At 8°C no viable cells. At 15°C reduction of CFU by 5 logs on turkey deli meats and in chocolate milk and by 3 logs on hot dogs	

Phage therapy of experimentally Salmonella-infected poultry and swine animals has been successful and significantly decreased Salmonella in major tissues such as ileum and cecal tonsils. Although these results have been mostly obtained within contrived laboratory conditions, the success of postslaughter phage employment, to lower the risk of cross-contamination, will only be determined after extending these studies to poultry farms.

Apart from one report in 2001, where the multivalent Felix01 phage was used, all other in vivo experiments performed recently were carried out using cocktails of two to six phages. Furthermore, Ma and colleagues (2008) have encapsulated phage Felix01 in a chitosan-coated Ca-alginate spheres [57] and have found in in vitro studies that this technique preserves phage viability upon exposure to acidic conditions; however, to our knowledge, these encapsulated phages have not yet been tested in vivo. Valuable outputs have been made in the last couple of years in terms of using phages for Salmonella biocontrol. Today, two phage-products are available: (1) BacWash from OmniLytics Inc. which received, in 2007, USDA's Food Safety and Inspection Services approval to be commercialized and applied as a mist, spray, or wash on live animals prior to slaughter; (2) BIOTECTOR S1 phage product from CheilJedang Corporation that is to be applied on animal feed to control Salmonella in poultry.

Unlike phage preharvest strategies on animals, several postharvest strategies have adopted the use of only one phage and not a cocktail. All Salmonella phages reported have been able to decrease the number of viable cells present on raw meats, processed and ready-to-eat foods, and fresh produce. Furthermore, the combined treatment of phage and Enterobacter asburiae, a strain exhibiting antagonistic activity against Salmonella, to control this pathogen on tomatoes, mung bean sprouts, and alfalfa seeds, represents a highly promising, chemical-free approach. However, in some settings phages were found to be readily immobilized by the food matrix and, although retaining infectivity, they lost the ability to diffuse and infect target cells [59].

Listeria monocytogenes is a gram-positive, motile, and facultative intracellular bacterium, that can grow under several food matrices and storing conditions (e.g., high salt levels, low pH, lack of oxygen, and low temperatures) [60]. Invasive infection by L. monocytogenes causes listeriosis and is transmitted to humans with 10^3 CFU/mL levels. It is often associated with contaminated minimally processed food such as ready-to-eat (RTE) products, poultry, and dairy products or related to cross-contamination after the heat treatment process of foods stored at low temperatures [61, 62]. Despite its low incidence, estimated to be in order of 2 of 10 reported cases per million per year in Europe [63], its pathogenicity causes a high mortality rate of approximately 255 deaths each year in the USA alone according to the Centers for Disease Control and Prevention (CDC) [1]. As so, contingency measures have led to the establishment of a limit for common RTE foods of 100 CFU per gram in EU and a zero tolerance policy in the USA.

Considering the sources of Listeria outbreaks, phage research has focused on postharvest applications (Table4).

Table 4: PostharvestListeriaphage applications

Year	Product	Phage(s)	Strategy	Main outcome	Refs
2002	Meat (ground beef)	Phage-nisin mixture	Applied on top	Phage-nisin mixture was effective in broth but not in buffer or on raw beef	[81]
2003	Fresh produce (melons, apples)	Phage cocktail (LM-103, LM-102) combined with nisin	Applied on top or sprayed	Phage caused a CFU reduction of 2.0 to 4.6 log in melons and only 0.4 log in apples. Phage + nisin reduced CFU by 5.7 (melon) and 2.3 (apple) log	[82]
2004	Fresh produce (honeydew melon)	Phage cocktail	Sprayed	Spraying melon pieces 0 h up to 1 h afterListeria challenge reduced CFU by 6.8 log units after 7 days of storage	[83]
2005	Processed food (red-smear soft cheese)	P100	Applied to surfaces during the rind washings	Reduction of CFU or complete eradication during the rind washings	[84]
2009	Processed food (cooked ham)	P100	Applied on top	Rapid 1 log reduction of CFU. 2 log reduction after 14 to 28 days of storage	[85]
2009	Fresh produce (ready-to-eat products)	A511, P100	Added to foods	In liquid foods, eradication of bacterial cells. On solid foods reduction of CFU by up to 5 log	[86]

2010	Meat (salmon fillet)	P100	Applied on top	Complete inhibition of growth at 4°C for 12 days, at 10°C for 8 days, and at 30°C for 4 days	[64]
2010	Meat (catfish fillets)	P100	Applied on top	Reduction of CFU by 1.4–2.0 log units at 4°C, 1.7–2.1 logs at 10°C, and 1.6–2.3 logs at 22°C	[87]
2011	Processed food (red-smear cheese)	A511	Applied on top	CFU counts dropped 3 logs after 22 days. Repeated application of A511 further delayed re-growth	[88]
2011	Processed food (ready-to-eat chicken)	FWLLM1	Added to foods	Reduction of CFU by 2.5 log at 30°C. At 5°C, regrowth was prevented over 21 days	[89]

The first two studies in Table 4 were carried out using a combination of phage and a nisin, a broad spectrum antibacteria peptide used during production to extend shelf life by suppressing gram-positive spoilage and pathogenic bacteria. While in ground beef the phage-nisin combination revealed to be ineffective, this strategy had a synergistic effect once added to melon and apple resulting in an improved reduction of Listeriacompared to phage or nisin alone. The efficacy of phage-nisin mixture was however significantly reduced in apples on the account of a decline of phage numbers possibly due to the low pH. Phage biocontrol should therefore be optimized separately for each food matrices under study. Four other studies have used phage P100, which was highly effective in inhibiting Listeria growth at storage temperatures for several days (see Table 4). Like in Campylobacter, only one article, by Soni and Nannapaneni (2010), on the efficacy of phages against biofilms was found for Listeria monocytogenes. These authors evaluated the ability of P100 against 21 L. monocytogenes strains belonging to 13 serotypes and found that P100 reduced by 3.5 to 5.4 log/cm^2 the viable cells present in stainless steel surfaces [64]. Nonetheless, studies carried out on a variety of experimentally contaminated meats, fresh produce, and processed food among others show that biocontrol is influenced by phage contact time and phage dose, regardless of higher or lower temperature. Despite this fact, the results conducted on RTE foods and meats using LMP-102 phage preparation (six-phage cocktail) allowed the commercialization of ListShield phage product from Intralytix. Also, LISTEX P100 phage-based-product (from EBI Food Safety) is being commercialized with a GRAS status (generally recognized as safe) to prevent Listeria contamination on food products and food processing facilities.

Staphylococcus aureus is a gram-positive bacterium, is considered to be a major threat to food safety [1, 90], and also is the most common agent of mastitis in dairy cows [91]. CDC estimates that staphylococcal food poisoning, which results from the consumption of foods containing sufficient amounts of one or more preformed enterotoxins [91, 92], is 242.148 cases annually in the USA [1]. The mechanisms described for the contamination of foods with S. aureus can be of animal or human origin, such as due to the infection and colonization of livestock or farm workers and even due to the human handling of the food products [93]. S. aureus can cause toxin-mediated diseases within 1 to 6 h after consumption of contaminated foods; nevertheless the symptoms are usually mild and most people recover within 1–3 days [94]. The annual estimated loss worldwide due to mastitis in adult dairy cows is of 35 billion US dollars [95]. Phage research has focused on the treatment of mastitis in lactating dairy cows and mostly in dairy food products (Table 5).

Table 5: Preharvest and Postharvest Staphylococcus aureus phage applications

Year	Animal/product	Phage(s)	Strategy	Main outcome	**Refs**
Preharvest application					
2005	Ruminant (dairy cattle)	K, CS1, DW2	Syringe applied into the teat sinus	No detectable increase in somatic cell counts in milk	[96]

2006	Ruminant (lactating dairy cattle)	K	Intramammary infusions	Cure rate comparable in phage-treated and saline-treated quarters. No large increase of somatic cell count in the milk when phage was infused into quarters with S. aureus infection	[97]
Postharvest application					
2005	Raw and ready-to-eat foods (milk products and derivatives)	K	Added to milk	Phages adsorption was reduced in raw milk and replication inhibited	[99]
2006	Raw food (raw milk whey)	K	Added to raw milk whey	Phage attachment and lysis inhibition due to adsorption of whey proteins to the S. aureus cell.	[98]
2007	Processes food (milk curd)	Cocktail (Φ88 and Φ35)	Added to pasteurized whole milk	S. aureus not detectable after 4 h at 25°C in acid curd and total clearance within 1 h of incubation at 30°C in renneted curd. The addition of the phage cocktail to milk prior to acid and enzymatic curd manufacture eliminates S. aureus by up to 6 log units	[100]
2008	Ready-to-eat foods (pasteurized milk)	Cocktail (Φ35, Φ88) with nisin	Applied to foods and mixed in milk	Nisin-phages application decreased S. aureus by 1 log unit more than in each antimicrobial agent alone (24 h at 37°C). Nisin-resistant phenotypes acquired resistance to phage, but phage resistance did not necessarily confer nisin resistance	[101]

2012	Ready-to-eat foods (pasteurized milk)	Cocktail (philPLA35, philPLA88 with high hydrostatic pressure (HHP)	Applied to foods and added to milk	Combination of HHP and phage resulted in S. aureus elimination within the 48 h regardless of the initial contamination level (1×10^6 or 1×10^4 CFU per mL)	[102]
2012	Ready-to-eat foods (cheese)	vB_SauS-phi-IPLA35, vB_SauS-phi-SauS-IPLA88	Added to pasteurized milk vat	Phage cocktail led to undetectable limits of S. aureus after 6 h in fresh cheese and continuous reductions in hard cheese. In curd a reduction of 4.64 log CFU per g was obtained compared with control	[103]

The experimental work of phage therapy on dairy cattle to control S. aureus on teats shows that, although there was no increase in somatic cell counts in milk samples indicating that the phages did not irritate the animal [96, 97], there were no statistically different outcomes between phage-treated and placebo groups [97]. This suggests that the use of phage as teat washes or sanitizers will not reduce the incidence of S. aureusmostly due to the barriers to phage-mediated bacteria lysis which are present in the bovine mammary gland. The suggested reasons for the inefficacy of phage K are the degradation or inactivation of the phage particles within the gland and the inhibitory effects of raw milk [97]. Also, it has been reported that milk whey prevents phages from reaching their host cell surface [98] due to the agglutination of S. aureus cells upon contact with raw whey [98, 99]. However, the work by García et al. [100] on the use of phages in curd manufacturing processes provides evidence that phages are stable and active during enzymatic curd formation suggesting that pH is, in fact, the most crucial inactivation factor. Furthermore, these authors postulated that the activity in milk of the phages used in their work was due to their dairy origin.

Postharvest applications in pasteurized milk show that the use of combined phage treatments with nisin and high hydrostatic pressure could synergistically be used to reduce staphylococcal contamination compared to each treatment alone [101, 102]. Nevertheless, care should be taken in regard to combined phage-nisin application since nisin-adapted strains can seriously compromise phage activity [101]. Inactivation of S. aureus has also been accomplished in both fresh and hard-type cheese using a phage cocktail during cheese manufacturing [103]. While for fresh cheese staphylococcal cells could not be detectable until the end of the curdling process (after 24 h), in hard cheeses, the presence of staphylococcal strains continuously dropped to 1.24 log CFU per gram at the end of ripening [103] providing evidence that phages can be successfully used to control S. aureus in dairy food products.

Today, there are phage-based diagnostic tools available to detect S. aureus, including antibiotic resistant (MRSA) and susceptible (MSSA) strains (Microphage, Inc.).

Besides the five foodborne pathogens described above, several other foodborne pathogens are responsible for illnesses, hospitalizations, and deaths, such as Clostridium spp., Shigella spp., and Vibrio spp. Nonetheless, these are not discussed further as to date there is a limited number of articles on the use of phage against these pathogens.

CONCLUDING REMARKS AND FUTURE PERSPECTIVES

Soon after the discovery of phages, several studies produced negative results frequently associated with their inappropriate usage, including administration to treat viral and unknown agent diseases. Now, when once more the potential of phages is being evaluated mostly due to the increasing emergence of multiresistant strains, it is important not to repeat the same mistakes. Furthermore, the adoption of phage administration and delivery routes, control procedures, concentrations, and timings of application among other parameters based on results found in the literature must be done with caution as the efficacy of each phage or cocktail of phages is highly dependent on the phage-host systems in study. Furthermore, sampling of experimentally contaminated animals and foods and their treatment/control with phage should mimic as much as possible the conditions found at preharvest and postharvest (slaughterhouse, processing, and at retail) stages. For example, to estimate the input of the slaughterhouse into the process chain, samples should be collected before and after cooling. Ready-to-eat food usually undergoes several processing stages, and thus it may be reasonable to verify the efficacy of phage after the process where phages were applied or, at least, at the end of the processing stage. Nevertheless, it is well evident that even after phage treatment, changes in food items still occur during transport and storage. Future studies should also focus on improving the general understanding of the mechanisms of phage resistance acquired by the hosts and the rate of elimination from the animal body. In addition, further evaluation of phage for biosanitation purposes is still not well documented for the pathogens listed above.

Although the results of the above-described studies appear to be encouraging, they should be interpreted cautiously. For instance, some phage studies have proven that phages are inefficient in reducing their host, such as phage delivery on foods and drinking water for controlling E. coli 0157 : H7 in poultry and ruminants, and reduced efficacy of some of the phages reported. Also, the emergence of phage-resistant phenotypes is reported by several authors; however this has not significantly affected the results of the phage trials on animals and can be managed with using phages which target those resistant phenotypes. Nonetheless, summarizing briefly the results described recently: (i) phage therapy is able to reduce foodborne pathogen levels in animals and consequently control the pathogen load on entry at the slaughterhouses; (ii) the strategies applied for phage biocontrol of pathogens in foods reduce significantly the levels in a variety of products and seem to be a promising alternative to traditional food safety and preservation measures; (iii) phage use in agricultural settings is as efficient or more than the conventionally used agents to control the growth of plant-based bacterial pathogens. However, the future of phages in food safety is further dependent on the regulatory agencies that still display uneasiness with using phages, mostly due to a scarcity of strong scientific evidence generated through fully controlled clinical trials under the supervision of ethical committees and in compliance with the highest regulatory standards of leading Western jurisdictions [90]. Also, there is a need to educate farmers, producers, and the general public about the advantages of their use.

AUTHORS' CONTRIBUTION

S. M. Sillankorva and H. Oliveira equally contributed to the paper.

ACKNOWLEDGMENTS

This work was supported by a Grant from the Portuguese Foundation for Science and Technology in the scope of the Projects PTDC/AGR-ALI/100492/2008. H. Oliveira and S. M. Sillankorva acknowledge the FCT Grants SFRH/BD/63734/2009 and SFRH/BPD/48803/2008, respectively.

REFERENCES

1. CDC, CDC Estimates of Foodborne Illness in the United States, Center for Disease Control and Prevention (CDC), 2011.
2. J. Rocourt, G. Moy, K. Vierk, and J. Schlundt, The Present State of Foodborne Disease in OECD Countries, 2003.
3. J. Maukonen, J. Mättö, G. Wirtanen, L. Raaska, T. Mattila-Sandholm, and M. Saarela, "Methodologies for the characterization of microbes in industrial environments: a review," Journal of Industrial Microbiology and Biotechnology, vol. 30, no. 6, pp. 327–356, 2003.
4. J. T. Holah, J. H. Taylor, D. J. Dawson, and K. E. Hall, "Biocide use in the food industry and the disinfectant resistance of persistent strains of Listeria monocytogenes and Escherichia coli," Journal of Applied Microbiology, vol. 92, pp. 111S–120S, 2002.
5. K. Lewis, "Multidrug tolerance of biofilms and persister cells," Current Topics in Microbiology and Immunology, vol. 322, pp. 107–131, 2008.
6. R. J. Atterbury, P. L. Connerton, C. E. R. Dodd, C. E. D. Rees, and I. F. Connerton, "Application of host-specific bacteriophages to the surface of chicken skin leads to a reduction in recovery of Campylobacter jejuni," Applied and Environmental Microbiology, vol. 69, no. 10, pp. 6302–6306, 2003.

7. F. C. Hsu, Y. S. C. Shieh, and M. D. Sobsey, "Enteric bacteriophages as potential fecal indicators in ground beef and poultry meat," Journal of Food Protection, vol. 65, no. 1, pp. 93–99, 2002.
8. J. E. Kennedy, C. I. Wei, and J. L. Oblinger, "Distribution of coliphages in various foods," Journal of Food Protection, vol. 49, no. 12, pp. 944–951, 1986.
9. V. B. Suárez, A. Quiberoni, A. G. Binetti, and J. A. Reinheimer, "Thermophilic lactic acid bacteria phages isolated from Argentinian dairy industries," Journal of Food Protection, vol. 65, no. 10, pp. 1597–1604, 2002.
10. D. L. Croci, D. de Medici, C. Scalfaro et al., "Determination of enteroviruses, hepatitis A virus, bacteriophages and Escherichia coli in Adriatic Sea mussels," Journal of Applied Microbiology, vol. 88, no. 2, pp. 293–298, 2000.
11. H. Rosenquist, N. L. Nielsen, H. M. Sommer, B. Nørrung, and B. B. Christensen, "Quantitative risk assessment of human campylobacteriosis associated with thermophilic Campylobacter species in chickens," International Journal of Food Microbiology, vol. 83, no. 1, pp. 87–103, 2003.
12. L. Snyder and W. Champness, "Lytic bacteriophages: genetic analysis and use in transduction," inMolecular Genetics of Bacteria, L. S. A. W. Champness, Ed., pp. 293–305, ASM Press, Washington, DC, USA, 2007.
13. D. Goode, V. M. Allen, and P. A. Barrow, "Reduction of experimental Salmonella and Campylobacter contamination of chicken skin by application of lytic bacteriophages," Applied and Environmental Microbiology, vol. 69, no. 8, pp. 5032–5036, 2003.
14. G. G. Greer, "Psychrotrophic bacteriophages for beef spoilage pseudomonads," Journal of Food Protection, vol. 45, pp. 1318–1325, 1982.
15. G. G. Greer, "Effect of phage concentration, bacterial density, and temperature on phage control of beef spoilage," Journal of Food Science, vol. 53, pp. 1226–1227, 1988.
16. T. Bigwood, J. A. Hudson, C. Billington, G. V. Carey-Smith, and J. A. Heinemann, "Phage inactivation of foodborne pathogens on cooked and raw meat," Food Microbiology, vol. 25, no. 2, pp. 400–406, 2008.

17. E. R. S. E. Data, Foodborne Illness Cost Calculator: Salmonella, 2010.
18. G. G. Greer, "Bacteriophage control of foodborne bacteria," Journal of Food Protection, vol. 68, no. 5, pp. 1102–1111, 2005.
19. J. B. Kaper, "Enterohemorrhagic Escherichia coli," Current Opinion in Microbiology, vol. 1, no. 1, pp. 103–108, 1998.
20. J. B. Russell, F. Diez-Gonzalez, and G. N. Jarvis, "Symposium: farm health and safety invited review: effects of diet shifts on Escherichia coli in cattle," Journal of Dairy Science, vol. 83, no. 4, pp. 863–873, 2000.
21. J. M. Whichard, N. Sriranganathan, and F. W. Pierson, "Suppression of Salmonella growth by wild-type and large-plaque variants of bacteriophage Felix O1 in liquid culture and on chicken frankfurters,"Journal of Food Protection, vol. 66, no. 2, pp. 220–225, 2003.
22. W. E. Huff, G. R. Huff, N. C. Rath, J. M. Balog, and A. M. Donoghue, "Prevention of Escherichia coliinfection in broiler chickens with a bacteriophage aerosol spray," Poultry Science, vol. 81, no. 10, pp. 1486–1491, 2002.
23. W. E. Huff, G. R. Huff, N. C. Rath, J. M. Balog, and A. M. Donoghue, "Evaluation of aerosol spray and intramuscular injection of bacteriophage to treat an Escherichia coli respiratory infection," Poultry Science, vol. 82, no. 7, pp. 1108–1112, 2003.
24. W. E. Huff, G. R. Huff, N. C. Rath et al., "Prevention of Escherichia coli respiratory infection in broiler chickens with bacteriophage (SPR02)," Poultry Science, vol. 81, no. 4, pp. 437–441, 2002.
25. L. D. Goodridge and B. Bisha, "Phage-based biocontrol strategies to reduce foodborne pathogens in foods," Bacteriophage, vol. 1, no. 3, pp. 130–137, 2011.
26. H. W. Smith, M. B. Huggins, and K. M. Shaw, "Factors influencing the survival and multiplication of bacteriophages in calves and in their environment," Journal of General Microbiology, vol. 133, no. 5, pp. 1127–1135, 1987.
27. S. Hagens and M. J. Loessner, "Application of bacteriophages for detection and control of foodborne pathogens," Applied Microbiology and Biotechnology, vol. 76, no. 3, pp. 513–519, 2007.

28. K. Stanford, T. A. McAllister, Y. D. Niu et al., "Oral delivery systems for encapsulated bacteriophages targeted Escherichia coli o157: H7 in feedlot cattle," Journal of Food Protection, vol. 73, no. 7, pp. 1304–1312, 2010.
29. S. J. Bach, T. A. McAllister, D. M. Veira, V. P. J. Gannon, and R. A. Holley, "Effect of bacteriophage DC22 on Escherichia coli O157:H7 in an artificial rumen system (Rusitec) and inoculated sheep,"Animal Research, vol. 52, no. 2, pp. 89–101, 2003.
30. W. E. Huff, G. R. Huff, N. C. Rath, and A. M. Donoghue, "Evaluation of the influence of bacteriophage titer on the treatment of colibacillosis in broiler chickens," Poultry Science, vol. 85, no. 8, pp. 1373–1377, 2006.
31. R. R. Raya, P. Varey, R. A. Oot et al., "Isolation and characterization of a new T-even bacteriophage, CEV1, and determination of its potential to reduce Escherichia coli O157:H7 levels in sheep," Applied and Environmental Microbiology, vol. 72, no. 9, pp. 6405–6410, 2006.
32. H. Sheng, H. J. Knecht, I. T. Kudva, and C. J. Hovde, "Application of bacteriophages to control intestinal Escherichia coli O157:H7 levels in ruminants," Applied and Environmental Microbiology, vol. 72, no. 8, pp. 5359–5366, 2006.
33. E. A. Rozema, T. P. Stephens, S. J. Bach et al., "Oral and rectal administration of bacteriophages for control of Escherichia coli O157:H7 in feedlot cattle," Journal of Food Protection, vol. 72, no. 2, pp. 241–250, 2009.
34. A. Oliveira, R. Sereno, and J. Azeredo, "In vivo efficiency evaluation of a phage cocktail in controlling severe colibacillosis in confined conditions and experimental poultry houses," Veterinary Microbiology, vol. 146, no. 3-4, pp. 303–308, 2010.
35. L. Rivas, B. Coffey, O. McAuliffe et al., "In vivo and ex vivo evaluations of bacteriophages e11/2 and e4/1c for use in the control of Escherichia coli O157:H7," Applied and Environmental Microbiology, vol. 76, no. 21, pp. 7210–7216, 2010.
36. R. R. Raya, R. A. Oot, B. Moore-Maley et al., "Naturally resident and exogenously applied T4-like and T5-like bacteriophages can reduce Escherichia coli O157:H7 levels in sheep guts," Bacteriophage, vol. 1, no. 1, pp. 15–24, 2011.

37. G. O'Flynn, R. P. Ross, G. F. Fitzgerald, and A. Coffey, "Evaluation of a cocktail of three bacteriophages for biocontrol of Escherichia coli O157:H7," Applied and Environmental Microbiology, vol. 70, no. 6, pp. 3417–3424, 2004.
38. T. Abuladze, M. Li, M. Y. Menetrez, T. Dean, A. Senecal, and A. Sulakvelidze, "Bacteriophages reduce experimental contamination of hard surfaces, tomato, spinach, broccoli, and ground beef byEscherichia coli O157:H7," Applied and Environmental Microbiology, vol. 74, no. 20, pp. 6230–6238, 2008.
39. M. Sharma, J. R. Patel, W. S. Conway, S. Ferguson, and A. Sulakvelidze, "Effectiveness of bacteriophages in reducing Escherichia coli O157:H7 on fresh-cut cantaloupes and lettuce," Journal of Food Protection, vol. 72, no. 7, pp. 1481–1485, 2009.
40. S. Viazis, M. Akhtar, J. Feirtag, and F. Diez-Gonzalez, "Reduction of Escherichia coli O157:H7 viability on hard surfaces by treatment with a bacteriophage mixture," International Journal of Food Microbiology, vol. 145, no. 1, pp. 37–42, 2011.
41. J. Patel, M. Sharma, P. Millner, T. Calaway, and M. Singh, "Inactivation of Escherichia coli O157:H7 attached to spinach harvester blade using bacteriophage," Foodborne Pathogens and Disease, vol. 8, no. 4, pp. 541–546, 2011.
42. Y. D. Niu, Y. Xu, T. A. McAllister et al., "Comparison of fecal versus rectoanal mucosal swab sampling for detecting Escherichia coli O157:H7 in experimentally inoculated cattle used in assessing bacteriophage as a mitigation strategy," Journal of Food Protection, vol. 71, no. 4, pp. 691–698, 2008.
43. J. E. Moore, D. Corcoran, J. S. Dooley, et al., "Campylobacter," Veterinary Research, vol. 36, no. 3, pp. 351–382, 2005.
44. T. Humphrey, S. O'Brien, and M. Madsen, "Campylobacters as zoonotic pathogens: a food production perspective," International Journal of Food Microbiology, vol. 117, no. 3, pp. 237–257, 2007.
45. S. J. Evans and A. R. Sayers, "A longitudinal study of campylobacter infection of broiler flocks in Great Britain," Preventive Veterinary Medicine, vol. 46, no. 3, pp. 209–223, 2000.
46. S. P. Oliver, B. M. Jayarao, and R. A. Almeida, "Foodborne pathogens in milk and the dairy farm environment: food safety

and public health implications," Foodborne Pathogens and Disease, vol. 2, no. 2, pp. 115–129, 2005.

47. F. Reich, V. Atanassova, E. Haunhorst, and G. Klein, "The effects of Campylobacter numbers in caeca on the contamination of broiler carcasses with Campylobacter," International Journal of Food Microbiology, vol. 127, no. 1-2, pp. 116–120, 2008.

48. C. L. Carrillo, R. J. Atterbury, A. El-Shibiny et al., "Bacteriophage therapy to reduce Campylobacter jejuni colonization of broiler chickens," Applied and Environmental Microbiology, vol. 71, no. 11, pp. 6554–6563, 2005

49. A. El-Shibiny, A. Scott, A. Timms, Y. Metawea, P. Connerton, and I. Connerton, "Application of a group II campylobacter bacteriophage to reduce strains of Campylobacter jejuni and campylobacter coli colonizing broiler chickens," Journal of Food Protection, vol. 72, no. 4, pp. 733–740, 2009.

50. C. M. Carvalho, B. W. Gannon, D. E. Halfhide et al., "The in vivo efficacy of two administration routes of a phage cocktail to reduce numbers of Campylobacter coli and Campylobacter jejuni in chickens,"BMC Microbiology, vol. 10, article 232, 2010

51. A. E. Scott, A. R. Timms, P. L. Connerton, C. Loc Carrillo, K. Adzfa Radzum, and I. F. Connerton, "Genome dynamics of Campylobacter jejuni in response to bacteriophage predation," PLoS pathogens, vol. 3, no. 8, p. e119, 2007

52. T. Ica, V. Caner, O. Istanbullu et al., "Characterization of mono- and mixed-culture Campylobacter jejuni biofilms," Applied and Environmental Microbiology, vol. 78, no. 4, pp. 1033–1038, 2012.

53. P. Siringan, P. L. Connerton, R. J. H. Payne, and I. F. Connerton, "Bacteriophage-mediated dispersal ofCampylobacter jejuni biofilms," Applied and Environmental Microbiology, vol. 77, no. 10, pp. 3320–3326, 2011

54. F. I. F. R. A. B. O. Germany, The Return of the Germs, 2004.

55. EFSA, "The European Union summary report on trends and sources of zoonoses, zoonotic agents and food-borne outbreaks in 2009," The European Food Safety Authority Journal, vol. 9, no. 3, p. 2090, 2011.

56. G. Barbara, V. Stanghellini, C. Berti-Ceroni, et al., "Role of antibiotic therapy on long-term germ excretion in faeces and digestive symptoms after Salmonella infection," Alimentary Pharmacology and Therapeutics, vol. 14, no. 9, pp. 1127–1131, 2000.

57. Y. Ma, J. C. Pacan, Q. Wang et al., "Microencapsulation of bacteriophage felix o1 into chitosan-alginate microspheres for oral delivery," Applied and Environmental Microbiology, vol. 74, no. 15, pp. 4799–4805, 2008

58. J. A. Wagenaar, M. A. P. V. Bergen, M. A. Mueller, T. M. Wassenaar, and R. M. Carlton, "Phage therapy reduces Campylobacter jejuni colonization in broilers," Veterinary Microbiology, vol. 109, no. 3-4, pp. 275–283, 2005

59. S. Guenther, O. Herzig, L. Fieseler, J. Klumpp, and M. J. Loessner, "Biocontrol of Salmonella Typhimurium in RTE foods with the virulent bacteriophage FO1-E2," International Journal of Food Microbiology, vol. 154, no. 1-2, pp. 66–72, 2012.

60. H. P. R. Seeliger and F. Jones, "Listeria," in Bergey's Manual of Systematic Bacteriology, B. Williams and M. D. Wilkins, Eds., pp. 1235–1245, 1986.

61. Y. C. Chan and M. Wiedmann, "Physiology and genetics of Listeria monocytogenes survival and growth at cold temperatures," Critical Reviews in Food Science and Nutrition, vol. 49, no. 3, pp. 237–253, 2009.

62. R. B. Tompkin, "Control of Listeria monocytogenes in the food-processing environment," Journal of Food Protection, vol. 65, no. 4, pp. 709–725, 2002.

63. H. de Valk, C. Jacquet, V. Goulet et al., "Surveillance of listeria infections in Europe," Euro Surveillance, vol. 10, no. 10, pp. 251–255, 2005.

64. K. A. Soni and R. Nannapaneni, "Removal of Listeria monocytogenes biofilms with bacteriophage P100," Journal of Food Protection, vol. 73, no. 8, pp. 1519–1524, 2010.

65. I. B. Sklar and R. D. Joerger, "Attempts to utilize bacteriophage to combat salmonella enterica serovar enteritidis infection in chickens," Journal of Food Safety, vol. 21, no. 1, pp. 15–29, 2001.

66. N. Lee and D. L. Harris, "The effect of bacteriophage treatment as a preharvest intervention strategy to reduce the rapid dissemination of Salmonella typhimurium in pigs," in Proceedings of the American Association of Swine Veterinarians, pp. 555–557, 2001.
67. L. Fiorentin, N. D. Vieira, and W. Barioni Jr., "Oral treatment with bacteriophages reduces the concentration of Salmonella Enteritidis PT4 in caecal contents of broilers," Avian Pathology, vol. 34, no. 3, pp. 258–263, 2005.
68. H. Toro, S. B. Price, S. McKee et al., "Use of bacteriophages in combination with competitive exclusion to reduce Salmonella from infected chickens," Avian Diseases, vol. 49, no. 1, pp. 118–124, 2005.
69. R. J. Atterbury, M. A. P. van Bergen, F. Ortiz et al., "Bacteriophage therapy to reduce Salmonella colonization of broiler chickens," Applied and Environmental Microbiology, vol. 73, no. 14, pp. 4543–4549, 2007
70. R. L. Andreatti Filho, J. P. Higgins, S. E. Higgins et al., "Ability of bacteriophages isolated from different sources to reduce Salmonella enterica serovar Enteritidis in vitro and in vivo," Poultry Science, vol. 86, no. 9, pp. 1904–1909, 2007.
71. C. Borie, I. Albala, P. Sànchez et al., "Bacteriophage treatment reduces Salmonella colonization of infected chickens," Avian Diseases, vol. 52, no. 1, pp. 64–67, 2008.
72. S. K. Wall, J. Zhang, M. H. Rostagno, and P. D. Ebner, "Phage therapy to reduce preprocessing Salmonella infections in market-weight swine," Applied and Environmental Microbiology, vol. 76, no. 1, pp. 48–53, 2010
73. T. R. Callaway, R. C. Anderson, T. S. Edrington et al., "What are we doing about Escherichia coliO157:H7 in cattle?" Journal of Animal Science, vol. 82, pp. E93–E99, 2004.
74. T.-H. Lim, D.-H. Lee, Y.-N. Lee et al., "Efficacy of bacteriophage therapy on horizontal transmission of Salmonella Gallinarum on commercial layer chickens," Avian Diseases, vol. 55, no. 3, pp. 435–438, 2011.
75. R. Modi, Y. Hirvi, A. Hill, and M. W. Griffiths, "Effect of phage on survival of Salmonella Enteritidis during manufacture and storage

of Cheddar cheese made from raw and pasteurized milk," Journal of Food Protection, vol. 64, no. 7, pp. 927–933, 2001.

76. B. Leverentz, W. S. Conway, Z. Alavidze et al., "Examination of bacteriophage as a biocontrol method for Salmonella on fresh-cut fruit: a model study," Journal of Food Protection, vol. 64, no. 8, pp. 1116–1121, 2001.
77. S. Pao, S. P. Randolph, E. W. Westbrook, and H. Shen, "Use of bacteriophages to control Salmonella in experimentally contaminated sprout seeds," Journal of Food Science, vol. 69, no. 5, pp. M127–M130, 2004.
78. J. P. Higgins, S. E. Higgins, K. L. Guenther et al., "Use of a specific bacteriophage treatment to reduce Salmonella in poultry products," Poultry Science, vol. 84, no. 7, pp. 1141–1145, 2005.
79. J. Ye, M. Kostrzynska, K. Dunfield, and K. Warriner, "Evaluation of a biocontrol preparation consisting of enterobacter asburiae JX1 and a lytic bacteriophage cocktail to suppress the growth of salmonella javiana associated with tomatoes," Journal of Food Protection, vol. 72, no. 11, pp. 2284–2292, 2009.
80. S. P. T. Hooton, R. J. Atterbury, and I. F. Connerton, "Application of a bacteriophage cocktail to reduce Salmonella Typhimurium U288 contamination on pig skin," International Journal of Food Microbiology, vol. 151, no. 2, pp. 157–163, 2011
81. G. A. Dykes and S. M. Moorhead, "Combined antimicrobial effect of nisin and a listeriophage againstListeria monocytogenes in broth but not in buffer or on raw beef," International Journal of Food Microbiology, vol. 73, no. 1, pp. 71–81, 2002.
82. B. Leverentz, W. S. Conway, M. J. Camp et al., "Biocontrol of Listeria monocytogenes on fresh-cut produce by treatment with lytic bacteriophages and a bacteriocin," Applied and Environmental Microbiology, vol. 69, no. 8, pp. 4519–4526, 2003.
83. B. Leverentz, W. S. Conway, W. Janisiewicz, and M. J. Camp, "Optimizing concentration and timing of a phage spray application to reduce Listeria monocytogenes on honeydew melon tissue," Journal of Food Protection, vol. 67, no. 8, pp. 1682–1686, 2004.
84. R. M. Carlton, W. H. Noordman, B. Biswas, E. D. de Meester, and M. J. Loessner, "Bacteriophage P100 for control of Listeria

monocytogenes in foods: genome sequence, bioinformatic analyses, oral toxicity study, and application," Regulatory Toxicology and Pharmacology, vol. 43, no. 3, pp. 301–312, 2005.

85. A. Holck and J. Berg, "Inhibition of Listeria monocytogenes in cooked ham by virulent bacteriophages and protective cultures," Applied and Environmental Microbiology, vol. 75, no. 21, pp. 6944–6946, 2009.

86. S. Guenther, D. Huwyler, S. Richard, and M. J. Loessner, "Virulent bacteriophage for efficient biocontrol of Listeria monocytogenes in ready-to-eat foods," Applied and Environmental Microbiology, vol. 75, no. 1, pp. 93–100, 2009.

87. K. A. Soni, R. Nannapaneni, and S. Hagens, "Reduction of Listeria monocytogenes on the surface of fresh channel catfish fillets by bacteriophage listex p100," Foodborne Pathogens and Disease, vol. 7, no. 4, pp. 427–434, 2010.

88. S. Guenther and M. J. Loessner, "Bacteriophage biocontrol of Listeria monocytogenes on soft ripened white mold and red-smear cheeses," Bacteriophage, vol. 1, no. 2, pp. 94–100, 2011

89. B. Bigot, W. J. Lee, L. McIntyre et al., "Control of Listeria monocytogenes growth in a ready-to-eat poultry product using a bacteriophage," Food Microbiology, vol. 28, no. 8, pp. 1448–1452, 2011.

90. EFSA, "The use and mode of action of bacteriophages in food production. Scientific opinion of the panel on biological hazards," The European Food Safety Authority Journal, vol. 1076, pp. 1–26, 2009.

91. Y. Le Loir, F. Baron, and M. Gautier, "Staphylococcus aureus and food poisoning," Genetics and Molecular Research, vol. 2, no. 1, pp. 63–76, 2003.

92. M. A. Argudín, M. C. Mendoza, and M. R. Rodicio, "Food poisoning and Staphylococcus aureusenterotoxins," Toxins, vol. 2, no. 7, pp. 1751–1773, 2010.

93. V. Spanu, C. Spanu, S. Virdis, F. Cossu, C. Scarano, and E. P. L. de Santis, "Virulence factors and genetic variability of Staphylococcus aureus strains isolated from raw sheep›s milk cheese," International Journal of Food Microbiology, vol. 153, no. 1-2, pp. 53–57, 2012.

94. "What You Need to Know about Foodborne Illness-Causing Organisms in the U.S," U.S. Food and Drug Administratio, 2012.
95. G. J. Wellenberg, W. H. M. van der Poel, and J. T. van Oirschot, "Viral infections and bovine mastitis: a review," Veterinary Microbiology, vol. 88, no. 1, pp. 27–45, 2002.
96. S. O›Flaherty, R. P. Ross, J. Flynn, W. J. Meaney, G. F. Fitzgerald, and A. Coffey, "Isolation and characterization of two anti-staphylococcal bacteriophages specific for pathogenic Staphylococcus aureus associated with bovine infections," Letters in Applied Microbiology, vol. 41, no. 6, pp. 482–486, 2005.
97. J. J. Gill, J. C. Pacan, M. E. Carson, K. E. Leslie, M. W. Griffiths, and P. M. Sabour, "Efficacy and pharmacokinetics of bacteriophage therapy in treatment of subclinical Staphylococcus aureus mastitis in lactating dairy cattle," Antimicrobial Agents and Chemotherapy, vol. 50, no. 9, pp. 2912–2918, 2006.
98. J. J. Gill, P. M. Sabour, K. E. Leslie, and M. W. Griffiths, "Bovine whey proteins inhibit the interaction ofStaphylococcus aureus and bacteriophage K," Journal of Applied Microbiology, vol. 101, no. 2, pp. 377–386, 2006.
99. S. O›Flaherty, A. Coffey, W. J. Meaney, G. F. Fitzgerald, and R. P. Ross, "Inhibition of bacteriophage K proliferation on Staphylococcus aureus in raw bovine milk," Letters in Applied Microbiology, vol. 41, no. 3, pp. 274–279, 2005.
100. P. García, C. Madera, B. Martínez, and A. Rodríguez, "Biocontrol of Staphylococcus aureus in curd manufacturing processes using bacteriophages," International Dairy Journal, vol. 17, no. 10, pp. 1232–1239, 2007.
101. B. Martínez, J. M. Obeso, A. Rodríguez, and P. García, "Nisin-bacteriophage crossresistance inStaphylococcus aureus," International Journal of Food Microbiology, vol. 122, no. 3, pp. 253–258, 2008.
102. R. Tabla, B. Martínez, J. E. Rebollo et al., "Bacteriophage performance against Staphylococcus aureus in milk is improved by high hydrostatic pressure treatments," International Journal of Food Microbiology, vol. 156, no. 3, pp. 209–213, 2012.
103. E. Bueno, P. García, B. Martínez, and A. Rodríguez, "Phage inactivation of Staphylococcus aureus in fresh and hard-type

cheeses," International Journal of Food Microbiology, vol. 158, no. 1, pp. 23–27, 2012..

Chapter 3

Agricultural Districts in the Italian Regions: Looking Toward 2020

Daniela Toccaceli

Department of Science for Economics and Business, University of Florence Via delle Pandette, 9 I-50127, Florence

ABSTRACT

Agricultural, quality agro-food, supply-chain and rural districts were introduced by Italian law in 2001. With reference to European structural policies, especially rural development policy, and to the districts' originality and relationship to the LEADER method and programmes, we wonder whether they can still form a useful part of the new programming period 2014-2020. The question is analysed in the wider context of international literature on socio-economic and economic policy, so as to stimulate broader interest, with a view to the future. A survey on national scale is presented, limited to the districts legally recognised by the Regions. A detailed historical reconstruction is given, offering a 'reconstructive reading' through which we clarify the actual nature of these districts, including their connections to clusters and industrial districts. We also discuss how to better link the theoretical and the legal frameworks. And we examine the connection between the basic economic mechanisms of the district theory and how the Regions have implemented them in reality. To do this, we summarised the overall conceptual framework into a few basic components for our desk research. This has led to development of a methodological definition of districts in agriculture.

Finally we give a concise analysis of the main outlines of the structural policies relating to this essential notion of 'district', in order to reach some conclusions of general validity.

BACKGROUND

The new approach for European structural policy for the period 2014–2020 gives the local scale a strategic role in implementing interventions in support of participatory local development. This is to be based upon multi-level governance (Committee of the Regions [2009]) and expansion of the LEADER method, especially in order to implement *community-led local development* (CLLD) and multi-fund strategies (European Commission's Directorates General AGRI, EMPL, MARE and REGIO[2013], European Union [2013]:357–359).

In Italy the Regions[a] are responsible for implementing these policies through their own programming and for integrating them using other policy tools. Many Regions had already recognised agricultural districts under the Italian law of 2001[b]. that empowered them to identify Districts in Agriculture[c]. (DAs) in the form of rural or quality agro-food districts.

[a]'Region' includes also the autonomous Provinces of Trento and Bolzano, which can legislate just like a Region.

[b]Legislative decree 18 May 2001 n. 228 « Orientation and modernization of the agricultural sector in accordance with Article 7 of Law 57 of March 5, 2001». Published on G.U. n.137 of 15/6/2001 Ord. Suppl. n.149. Art.13 «Rural and quality agro-food districts».

It is also named 'Orientation Law' in analogy with the French 'Loi d'orientation agricole'.

[c]In the following "DA" is used for generically indicating each type of district in agriculture.

At the time of its introduction, that law aimed to boost agricultural modernisation, following the new deal promoted by Agenda 2000. Today, with the new programming cycle of European structural policies and common agricultural policy, the important question is, how useful are the DAs and how can they be inserted into the new regional programming framework in order to implement rural development programmes?

Despite many studies into regional experiences of DAs[d], organic knowledge is still lacking about implementation of the law on a national level, which is still ongoing. Indeed, researching on a national scale has proved very complex for three reasons: i) DAs have become very different "things" in each regional interpretation, so it is difficult to make comparisons among them; ii) DAs are still an open chapter of economic and agricultural research, because of the particular theoretical framework that has been adopted in Italy; iii) the theoretical and legal frameworks are still not fully integrated (Musotti [2001b]).

[d]A not exhaustive list of studies on regional case studies includes: Pacciani ([2003]), Regione and Agenzia Regionale Servizi Sviluppo Agricolo ([1999]); Saraceno ([1992]);Casati et al. ([1999]);[Pacciani (2002b)]; Pacciani ([2003]); Belletti and Marescotti ([2004]); Conti ([2005]); Aimone ([2006]); Toccaceli ([2006]); CNEL ([2007]); Fazio and Ricciardi ([2008]); Gulisano et al. ([2008]); Hoffmann ([2008]); Marangon ([2008a]); Marangon [2008b]); Baldi and Casati ([2009]); Belletti and Marescotti ([2010]); Contò et al. ([2010]); Petrovich and Manzoni ([2010]); Caré et al. ([2010]); Contò and La Sala ([2011]); Garofoli ([2011]); Maizza ([2010]); Montresor et al. ([2011]); Tarangioli ([2008]); Tarangioli ([2011]); Spampinato and Timpanaro ([2012]).

We have addressed these difficulties by reconstructing the legislative situation stratified over a twenty-year period – the 'reconstructive reading' according to Albisinni ([2010]) - and by using a fresh approach. Our research aims to create the conditions for comparing experiences which have arisen in concurrence with very different regulatory frameworks and political contexts. Our article also tries to recreate an updated geographical mapping of the phenomenon at national level and seeks to outline the main economic attributes assumed by DAs. On the basis of our findings we offer our initial thoughts on the potential and weaknesses of future use of DAs in the coming programming period 2014–2020.

Although this article focuses on Italian agricultural, quality agro-food, supply-chain and rural districts, the issue is placed in the context of international socio-economic literature on the subject. The research may thereby be of more general interest, firstly because clusters and industrial districts are widespread at global level in the agricultural sector and also because of the original approach used for analysing regional policies.

Some different instruments, similar to DAs even if not derived from a 'district' concept, exist in other European countries, as mentioned below.

A Brief Literature Review of Clusters and Marshallian Industrial Districts

Agglomerations of companies closely interconnected, specialised and co-located in a specific place have been the subject of empirical observations and researches over many years and in many countries. The earlier Marshallian Industrial District (MIDs) notion originated what is nowadays considered the core of the theoretical framework for explaining their persistence and diffusion: external economies and "industrial atmosphere" (Becattini et al. [2009]). Though they have changed over time, they have been noted because of their success as well as, sometimes, for the excellent reputation of some product-place associations and specialisations, for example the "Prosciutto di Parma".

According to Becattini et al. ([2009]) and Porter and Ketels ([2009]), clusters and MIDs are neither synonyms nor concepts conceived in same scientific context or with reference to identical economic and societal environments, although they have "common roots".

Coming from a business strategy approach (Porter [1998]) to understand the determinants of competition, Porter and Ketels ([2009]) define clusters as "geographic agglomerations of companies, suppliers and service providers, and associated institutions in a particular field, linked by externalities and complementarities of various type" (2009:172). These authors consider them "by design, a broader concept and a more general one" (2009:182) than MIDs, which are instead considered as "one type of a cluster".

Conversely, Becattini's MIDs are defined as "socio-territorial entities characterised by the active presence of both a community of people and a population of firms in one naturally and historically bounded area with a dominant industrial activity" (Pyke et al. [1990], cit. in Porter and Ketels[2009]:172). Becattini adds, however, that "the population of firms and the related market institutions are only one, though central, part of it" (Becattini et al. [2009]: xviii) and consequently clusters are considered only the industrial core of a more complex district process.

For the purpose of this paper, it is worth underlining one significant difference and some essential, common key points.

While the concept of MIDs is strictly related to industrial manufacturing sectors as well as to a system of small-medium enterprises, the notion of cluster is "by design" recognised in any production or services sector (Porter [1998]). So agricultural clusters are identified in rural regions, are studied with the same methodological approach and are involved in identical considerations about their role in the value chain and about public policies (Porter [2003]; Porter et al. [2004]; Feser and Isserman A. [2009]). Instead, justifying the translation of the MID concept to agriculture has required great efforts on the part of Italian researchers, as discussed below.

In any case, both converge on these issues: i) the core of the theoretical framework rests on a few basic conceptual nodes: the proximity (spatial dimension) of companies and institutions, interrelated in vertical, horizontal and diagonal ways, allows for better organisation of the value chain (business dimension); more added value is provided by a special local mix of externalities resulting from (in MID terminology) industrial atmosphere or (in cluster terminology) business environment (socio-economic dimension); ii) both put at the core of their conceptual construction the strategic function of the relationships between companies, local institutions in a wide sense, and the governance mechanisms for better management of those relationships; at the same time both recognise the twofold presence of an internal (local) network as well as of an external network, each playing an important role in the district process (network and organising dimension); iii) both are positively correlated with local development processes and not only to the growth of the enterprises involved; iv) for both, researchers have addressed policy recommendations aimed at supporting their growth and the development of the place/region where they are located (policy dimension).

So, clusters and MIDs are recognised in many countries which have introduced specific policies.

In the discussion of the results we analyse how these basic components work together in the Italian regional patterns.

Cluster Policy Approaches

Porter ([2003]) survey on economic performance of U.S. regions claims that regional development policies have to be particularly attuned to traded clusters because of their capability to support higher wages but also because they drive local employment and affect regional economic performances. On the other hand, Feser and Isserman A. ([2009]) found that rural clusters play an important role in the national value chain. They maintained that public policies should support these clusters, not only to sustain rural development but also to achieve an overall goal of regional and national economic development. Still focusing on rural regions in the U.S., Porter et al. ([2004]) state the importance of "cluster thinking" in rural economic development, remarking that policy has to be set at local and regional level because of the heterogeneity of rural areas and that community-based planning is needed. A "cluster-based approach" is thought essential for involving the private sector, and an "overarching organisational structure" for economic development is deemed helpful to coordinate and drive the process. However, they note, the role of institutions in a wider sense in the process of economic development in rural areas requires further research to be better understood.

In many European countries cluster policies - including industrial district policies - have existed since the 90s. They have been addressed to systems of small and medium enterprises in various production sectors, as well as to networks in the cluster, either for developing them or sometimes for using a "cluster lens" to boost other instruments e.g. for innovation processes or regional development. An Oxford Research ([2008]) found it difficult to identify a clear pattern among the 27 countries investigated. The complex nature of a cluster entails that cluster policies are complex too, given their multiple dimensions and the multiple actors involved in the process of programming and implementing the policies. An important reason for this is that the structural policies of the European Union have come to be delivered through multi-level governance processes involving both local actors and, vertically, more distant subjects and administration levels. Meanwhile, principles of subsidiarity and proportionality[e.] were entering the political and legal systems and processes of decentralisation towards the most peripheral levels of administration were occurring in many countries, Italy included.\

[e]Just defined in Article 5 of the Treaty on European Union, both were issued since 1992 in a declaration of Edinburgh European Council-

Borràs and Tsagdis ([2008]) recognised five types of cluster policy, ranging from three top-down types that aim to create clusters from scratch ('creationist') or to support them without encompassing a bottom-up process ('narrow' and 'top-down'), to the two more complex types: the 'evolutionary' one that considers clusters tend to evolve in a cumulative and gradual way and policy is embedded in the network which shapes the direction of these changing dynamics, and the 'network' policy. The latter is based on multi-level governance because it recognises that the role of public authorities is inscribed into a relationship between private and public actors in the territory where they try to solve problems collectively. 'Network' policy is confident not only regarding success by public authorities but also the strength of the network.

And Other Related Policies for Rural Areas in Europe

In European rural areas some other policies have been implemented that involve the concepts of cluster and districts, in their approaches based on the concept of network, territorial value and governance, and their focus on innovation in the rural environment. In this paper we consider only policies for qualifying agriculture and agro-food products and on LEADER[f.] programmes and methods. Although not cluster-specific, these policies have played a specific role in the integrated mix of main policies implemented in those areas, especially with reference to the rural districts and the quality agro-food districts which have developed in Italy, as further discussed below.

[f]LEADER is the acronym for: Liason Entre Actions de Développement de l'Économie Rurale, that is "Links between actions for the development of the rural economy".

Stemming from the 1988 reform of structural funds, several LEADER programmes have developed continuously, undergoing various adjustments and growing in importance in the scope of rural policies, until they became integrated in rural development programmes in the past and in the current period. This was done without changing the basic methodological principle to which the specific 'LEADER added value' is attributed: bottom-up approach, territorial and integrated approach to rural development, public-private partnership in local action groups (LAGs), territorial networking and transnational cooperation to stimulate rural innovation processes. In other words, LEADER is a "network policy" in the sense that it looks at different types of networks with a multi-purpose intent. Firstly, at the 'social' network, which addresses the intangible factor of development through reinforcing the sense of local identity of the rural population, as well as the external image of the territory. This boosts marketing of local products (Ray [1998]), because in LEADER a thread connects the sociological entity 'community' and the geographical entity 'territory' "that can be linked with an identifiable cultural heritage and identity" (Lee et al. [2005]:275), as can the socio-economic entity 'social capital'. Secondly, the 'horizontal' network, as Murdoch ([2000]) put it, "linked to more general and also non-agricultural processes", which aims at integrated development of rural economy in a defined area. Thirdly, the 'vertical' network, which links actors of a specific area because of their involvement in the agro-food sector (Murdoch[2000]).

Governance is therefore a fundamental process for the success of this policy, which has often attracted criticism about its capability to concretely and effectively achieve the prefigured added value (European Court of Auditors [2010]).

For several years voluntary certification schemes have been in use in Europe[g] offering protected status to foodstuffs with a certified local identity. This is a mechanism for adding value to such foods, thereby aiding local economies and supporting to some extent rural development (Tregear et al. [2007]). In these policies too, it is the role of network and governance that is interesting for our purposes. Regarding the designation of origin supply chains (*filière*) Perrier-Cornet and Sylvander ([2000]) focus, firstly, on the development of collective strategies through coordination by *filière* actors in various ways. The aim of these strategies is to gain a collective, competitive advantage linked to a specific place and shared ways of working and traditions. This advantage depends on quality territorial revenue (Pecqueur [2001]; Mollard [2001]). Secondly, they discuss the more complex territorial strategies developed using various mixes of sectorial and territorial governance related to different degrees of territorial anchorage of the supply chains.

[g]Three EU schemes promote and protect names of quality agricultural products and foodstuffs: Protected Designation of Origin - PDO: covers agricultural products and foodstuffs which are produced, processed and prepared in a given geographical area using recognised know-how. Protected Geographical Indication - PGI: covers agricultural products and foodstuffs closely linked to the geographical area. At least one of the stages of production, processing or preparation takes place in the area. Traditional Speciality Guaranteed - TSG: highlights traditional character, either in the composition or means of production.

The overlap seems self-evident between the two types of European policies directed at farms and agro-food business supply chains, which sometimes may be spatially concentrated in local systems, if not coinciding with a recognised district. Sforzi and Mancini ([2012]), however, suggest it is worth bearing in mind the difference between a 'simple' localised agro-food system (Courlet and Pecqueur [1992]) and a district, which implies a socioeconomic process (local community involvement) that is not present in the former.

Finally, worth mentioning is comparative research (Pacciani and Toccaceli [2014]; Toccaceli[2014]) into some case studies in several European member states, seeking to identify methodological analogies between DAs and other instruments developed to manage rural governance. This found considerable similarities with the French *contrat de Pays* (Toccaceli[2012]:28) and the functioning of the *collectivités territoriales* (Rouger [2014]), as well as with other systems in Belgium (Collignon [2014]) and LEADER-based experiences in Spain (Gallego Moreno [2014]).

DAs in the Debate by Italian Agricultural Economists

In Italy the MID notion began to be applied to the agricultural sphere from the end of the '80s, as a result of contamination between industrial and agricultural economists (Angeli [2000]), in a context both of visible change in agricultural structure and of progressive differentiation of production in the countryside. Meanwhile, European rural development policy started to be formulated with the European Commission's communication "The Future of Rural Society" (Commission of the European Communities [1988]). In Italy, Iacoponi ([1990]) began[h.] by putting the question as simply an extension to agriculture of the Marshallian district concept re-elaborated by Becattini ([1987], [1989]) for Italian industrial districts.

[h]Towards the end of the '80s the contributions (1987 and 1989) where Becattini systematizes the studies conducted on the industrial district of Prato and formulate the paradigm of the industrial district were published. At that time, some studies were carrying out on the contribution of agriculture to the determination of the pre-conditions for the development of an industrial district (Cecchi [1988]; Cianferoni [1990]; Cecchi et al. [1991]).

However, the debate soon became an occasion for rethinking (and disputing) the comparison between "two opposite economic paradigms*[i.]" (Iacoponi [1990]; Cecchi [1992]; Becattini [2000] and Becattini [2001]; Iacoponi [2001]): the neoclassical and the districtual. As a result, the "systemic" stream of economic agricultural thought was revisited, deriving from the "territorial" approach (De Benedictis [2000]; De Filippis [2000], Fabiani [2000]; Angeli [2000]; Musotti [2001a]) observing that "the reality was beginning to escape the theory*" (Favia [2000]). But only some agricultural economists were willing to welcome the 'district' paradigm, so limited but essential contributions referring to the notion of agricultural district were produced.

[i]Here and below the symbol "*" indicates our translation from italian.

Iacoponi ([2001]) considered the paradigm of the industrial district proposed by Becattini to be composed of two core concepts, interacting dynamically in the local district system: the production-organisational (typically Marshallian) and the socio-economic (typically Becattinian). Such an artificial separation is useful in neatly clarifying the different notions of districts in agriculture: the agro-industrial district and the agricultural district, both related to the production-organisational approach (Iacoponi [1990], Cecchi [1992]); and the rural district that better reflects the socio-economic approach (Romano [2000]; Iacoponi [2002]).

The Notions of Agricultural District and Agro-Industrial District

Iacoponi based his assessment of the compatibility between industrial and agro-industrial districts on a theoretical elaboration aimed at reconnecting the company-oriented economic approach typical of economic agrarian tradition, with the "contribution of the modern theory of the firm and developing it in such a direction as to include production systems organised at the local level*" (1990:711). According to an integrated vision of agriculture (Casati [1988]) he took into account the overall production structures of agriculture and observed that only few farms are autonomous from an organisational point of view (1990). All the others are integrated either into the general economic system of the territory, or into the agro-industrial system. In the latter case, depending on the agribusiness structure, integration can be either at the level of the local system of enterprises (agricultural district), or at the level of product market (supply-chain integration), or at the level of the local system of firms and product markets (agro-industrial district). As an example of agro-industrial district Iacoponi cited the case of Parmigiano Reggiano.

Cecchi accepted Iacoponi's idea and elaborated it with more attention to the farms' territorial value. He therefore argued that: "The designation of agricultural district has to be found in the centrality of agricultural production: that is, it must be able to identify local agriculture as the "engine" of a district; so even in the presence of agro-industrial activities, the latter exists and thrives because it represents a processing of local agricultural products*" (1992:96). As an example of agricultural district he cited the Chianti wine district. He then goes on to provide a definition of agro-industrial district, which is when "the main distinguishing element is represented by the fact that the processing industry will use mainly agricultural products from outside the district itself*" (1992:97). The example referred to is the Po milk chain, which (evolving from an agricultural district) transforms milk mainly from external sources.

The Notion of Rural District

The rural district is analysed in relation to the various forms of socio-economic organisation that can characterise local systems in the countryside (Montresor [2000]; Romano [2000]; [Forleo 2000]). This covers not only areas where agricultural production is prevalent, but also local systems of rural development or peri-urban areas where production differentiation has not occurred (Cecchi [2000]) or agro-districts and, lastly, rural districts. Cecchi described patterns of farm integration into the overall system of the territory, where the development of economic organisation solutions has enhanced the contextual knowledge and the capabilities of cooperation and social integration of local actors. He therefore distinguished "the districts of the areas of agricultural specialisation from those of rural local systems*" (2000:125).

Consequently, rather than connect the rural to the Marshallian district, according to the basic principles set out by Becattini ([2000]), Cecchi uses the "district approach" to separate social processes from economic ones. He thus highlights their specific contribution to and support of production differentiation of the countryside and sees the consolidation of the rural economy as having a plurality of economic activities rather than a predominance of agriculture. On this basis, he asserts that the possibility of identifying rural districts rather than agro-industrial districts is "more likely", although it is not easy to "perceive" a rural district because of the greater "physical distance between people and businesses" (2000:128). He thereby underlines how the spatial concept of proximity is changeable in the rural context and very different from the industrial one, especially in relation to MIDs.

Iacoponi, on the other hand, states that "the hypothesis that rural development is a local, sustainable, integrated and endogenous development leads us to the concept of the rural district, which is derived from the concept of industrial district (Becattini [1987]; Bellandi [1987]) and agro-industrial district (Iacoponi [1990]; Fanfani and Montresor [1998]; Cecchi [1992]; Favia [1992]).*" (2002:74). Following Becattini's approach, he distinguishes firmly the notion of rural district from the weaker notion of a geographically significant network, because the first "is a strengthening and intensification of relations historically settled in the territory thanks to agriculture and activities related to them*" (2000:76). In fact, he notes that the main character of a rural district is the ability to provide districtual synergies for the economic activities of small and medium enterprises which have a sporadic presence in the territory, where "the community market and the atmosphere of the rural district can perform the miracle of integrating the 'missing branches' of the local economy.*" (2002:75–76), So Iacoponi highlights the basic socio-economic engine of the rural district.

Both scholars wrote taking as reference the case of the rural district of Maremma (Pacciani [1997]; Pacciani [2002a] and Pacciani [2003]; Belletti and Marescotti [2004] and Belletti and Marescotti[2010]; Pacciani and Toccaceli [2010]; Toccaceli [2012]) promoted by Pacciani in 1996. Pacciani bypassed the scientific debate to enter the experimental and operational dimension oriented towards planning and implementation of the rural district, in its first prototypal form. His approach was independent of the idea of "transposing" the notion of industrial district to agriculture. Rather, he reworked and combined certain elements in the political and cultural debate of the '90s, including the gradual strengthening of European rural development policy and local development support (De Rita and Bonomi [1998]) to be pursued through specific instruments of negotiated planning (CNEL [1999]; De Meo and Nardone [2002]). Pacciani's contribution (Pacciani [2002b]) is particularly significant because he suggested a conceptual link between the rural district and LEADER, just as strong as (if not stronger than) the one with Marshallian industrial districts.

For the economists mentioned above, it is self-evident that the overlap of the concepts of local and rural development in the notion of rural district is strictly linked to all the new functions of agriculture (economic, environmental and socio-cultural). Later both Quality Agro-Food Districts (DAQ) and Rural Districts (DIRU) came to be considered a positive response to the rapid development of globalisation (Tinacci Mossello [2002]) which is pushing for the relocation of many industrial districts (Corò and Grandinetti [1999]). DAs, by contrast, as a "nemesis of history" (especially quality agro-food districts, see De Rosa and De Vincenzo [2002]:14), react by putting in place a strategy deeper rooted in the territory, conducted by enhancing typical products and multi-functionality of agriculture.

The Political Approach in Italy

Pacciani defines the rural district of Maremma "a complete implementation of the Cork Declaration*", introducing it into the agrarian-economic debate as a "development model and method of governance*" in rural areas (Pacciani [2003]:48). In this sense, the rural district was also taken into account by the government.

With his formulation, Pacciani offers a response to the policy issues raised by Iacoponi and Cecchi. Iacoponi ([1990]) believes that the Common Agricultural Policy (CAP) should strengthen existing districts and encourage the growth of those new ones. Cecchi considers that regulatory measures will be perceived differently, as if through "filters", in different places and this implies the existence of a local demand for policies that should find a "strong political mediation, implemented on the basis of territorial articulation" (1992:88).

The issue of how such mediation may legitimately be conducted, although noted, was left open while in practice the rural district as a "method of governance" was already providing some local responses.

In 2001 the economic debate was bypassed by government intervention placing the DAs in the context of legislation regarding modernisation of agriculture[j]. This regarded DAs as new tools for increasing competitiveness of the primary sector according to European guidelines for reviewing the CAP and rural development policy.

[j]See note b.

Because of its pragmatic aim, the government only partially followed indications arising from the scientific debate, rather considering pilot experiences autonomously occurring in Maremma and Piedmont (Toccaceli [2012]).

The government was particularly driven by the progressive intertwining of legislation on industrial districts and negotiated planning, whose applicability was extended to all sectors, including agriculture, thus opening the door to district legislation in that sector (Albisinni [2010], [2011a] and Albisinni [2011b]). In the legal rationale, in fact, the DAs – and particularly the rural districts - were also taking on a second, but not secondary, meaning as tools for supporting local development of rural areas.

The news was not welcomed unreservedly by a part of the scientific world. Although extensive preparatory work had been carried out, the final text does not appear to have fully taken it into account and Musotti ([2001b]) observed that this law would scarcely help in identifying and properly "recognising" the district situation.

The controversy highlighted a misconception that was implicit in national law and that was to become increasingly more evident in the regional laws. Economic theory had studied those districts that had formed spontaneously, therefore the law should enable their recognition. Instead, in a 'creationist' interpretation – well evident in several regional laws – the DAs were to become subjects to be set up for implementing policies at local level.

In this blurry legislative context, the Regions have made and are still making very different choices about timing, as well as about the definition and aims of DAs and the rules of implementation to be adopted.

The DAs in National and Regional Legislation

Because DAs refer to a composite regulatory framework that was created over a twenty-year period, a "reconstructive reading" (Albisinni [2010]) was needed to evaluate the complexity of the phenomenon and the political intentions behind it. Bearing in mind that interest in DAs in Italy ran parallel to the evolution of European rural development policy, the succession of national regulatory interventions might be seen as falling into three periods, corresponding to different political and economic approaches. (See Table 1).

Table 1: Evolutionary path of the European and Italian regulations referring to the DAs and related definitions

EU political references	Italian National laws		Legal notions	Definitions
C.I LEADER (1988–2006)	1991-2001	First law on industrial districts	**Industrial Districts**Law 317/91 art.36 para.1	*Local territorial areas with high concentration of small enterprises, with particular reference to relationship between presence of enterprises and resident population, as well as to production specialisation of enterprises as a whole.*
		Regulation of negotiated planning	Negotiated planning L. 104/1995	*Regulations agreed between public entities, or with competent public entity, and party or parties, public or private, for implementation of different interventions related to a unique development purpose that requires comprehensive assessment of activities pertaining to it.*
EU Rural development policy			1998 Del. CIPE	*Applicability of negotiated planning to rural, agricultural and fisheries districts*
		Second law on industrial districts	**Local productive systems (LPS)** Law 140/99 art.6 para.8	*Homogeneous productive contexts with high concentration of industrial enterprises and specialisation of business systems*
		The Orientation Law	**Rural districts**Legislative decree 228/2001 art.13 para.1	*Local production systems referred to in Article 36, paragraph 1, of Law of 5 October 1991 n. 317 [see first row above] and subsequent amendments with homogeneous identity from a historical and territorial point of view arising from integration between agricultural activities and other local activities, as well as production of goods or services of particular specificity, consistent with traditions and natural and territorial vocations.*
			Quality agro-food districts Legislative decree 228/2001 art.13 para.2	*Local production systems, including interregional, with significant economic presence and productive interrelationship, and interdependence of farms and agri-food enterprises, and one or more certified or protected products in accordance with applicable Community or national regulations, or from traditional or typical products.*

LEADER AXIS (2007–2013)	2002-2008	The Finance Law 2006	**Productive Districts**Law 266/2005 art.1 para.366	*Free combinations of enterprises articulated territorially and functionally, aiming at enhancing development of areas and sectors, improving effective organisation and production, according to principles of vertical and horizontal subsidiarity, also identifying ways of working with entrepreneurial associations.*
				Freely accessible by industrial companies, services, tourism and agriculture and fisheries.
	2008-2013	Business networks and network contracts	**Business network**L.133/2008 art.6bis	*Free aggregations of single production centres cohesive in unitary development of industrial policies, also in order to improve their presence in international markets*
Extension of LEADER method to ESI FUNDS	2014-2020	**What role for the DAs?**		

Toccaceli

Toccaceli *Agricultural and Food Economics* 2015 3:1 doi: 10.1186/s40100-014-0019-9

Table 2: DA typologies related to the main legal basis (Source: Toccaceli D. 2012:54)

	Local Productive System – (LPS)	Productive districts - (DP)
Rural and Quality Agro-Food districts (DRAQ)	DIRU DAQ	DP_RU DP_AQ DIF
Industrial Districts (DIND)	DINDA	DP_A

Toccaceli

Toccaceli *Agricultural and Food Economics* 2015 3:1 doi: 10.1186/s40100-014-0019-9

Table 3: DAs recognised by the Regions by main production sectors (Source: own elaboration)

		DAQ	DIRU	DP_RU	DP_AQ	DIF	**DP_A**	[redacted]	
Non specialized	Multi-product						1	8	9
	Basket products-territory	6	11	4	2				23
	Urban–rural		1	4					5
	Forestry rural		1						1
	Subtotal	6	13	8	2	0	1	8	38
Specialized	Agro-energy		1			1			2
	Food biotechnologies							2	2
	Coffe							1	1
	Cereals	1					1	1	3
	Confectionery						1		1
	Plant Nurseries	2	2			2	1		7
	Fish						2		2
	Milk and milk dairy products	1				1	2	1	5
	Fruit and vegetables	3				1	4	1	9
	Cork							1	1
	Wine	2					2	1	5
	Animal husbandry					3	3		6
	Subtotal	9	3	0	0	8	16	8	44
	Total	15	16	8	2	8	17	16	82

Toccaceli

Toccaceli *Agricultural and Food Economics* 2015 3:1 doi:10.1186/s40100-014-0019-9

The first period corresponds to the years between the first law on industrial districts in 1991 and the Law of Orientation of 2001. At this stage, the action of the government was inspired by the constant intertwining of the initial scientific debate on industrial districts and the possibility of their extension to agriculture, the progress of the negotiated planning experience (Adornato [2005]; Pacciani [2003]; Albisinni [2011a]) and the acknowledgment of some pilot experiments.

The early definition of industrial district (Table 1) well represents the basic components of the scientific notion in the spatial, business and social dimensions. It is limited to the industrial sector and only thanks to the formulation of the local production system notion did it pave the way for the extension of the district into different areas, such as rural, agro-food and fisheries. Similarly, in that period, negotiated planning instruments were being extended to cover agriculture and fisheries, services and tourism. The concept of negotiated planning explains the purpose of the first tools used by the DAs and allows us to appreciate the coherence of the two approaches.

At the same time, with negotiated planning, it became evident that the local approach required the establishment of a new kind of governance and specific tools to bring together different levels of public administration and public with private bodies. These tools must have the sole purpose of identifying specific solutions to sustain strategic projects supporting endogenous development and employment. In the wake of these measures, the Law of Orientation in agriculture introduced two distinct definitions for rural and quality agro-food districts, while maintaining a strong link with the basic concept of Local Production Systems (LPS), defined as *"the homogeneous productive contexts characterised by both a high concentration of industrial enterprises and the specialisation of business systems"*. These LPS *"characterised by homogeneous identity from a historical and territorial point of view, arising from integration between agricultural activities and other local activities, as well as the production of goods or services of particular specificity, consistent with traditions and natural and territorial vocations"* are qualified as Rural Districts (DIRU). The definition of Quality Agro-food Districts (DAQ), on the other hand, is given to those LPS *"even interregional, characterised by significant economic presence and production interrelationship, and by interdependence of farms and agri-food enterprises, and by one or more certified or protected products in compliance with applicable Community or national regulations, or by traditional or typical products".* Without contesting Musotti ([2001b]), one may observe that these definitions more or less mirror the main terms of the scientific debate.

In the second period (2002–2008) Italian law continued to mature intentions and purposes. The first decade of the adoption of districts and negotiated programming tools had responded to the objective of increasing employment by supporting endogenous local development processes. In the next period, attention reverted to focussing primarily on support for districtual firms in their processes of innovation, rationalisation of production, supply chain and internationalisation. So in the 2006 finance law the definition of 'district' was revised and reworked in terms of "free association of undertakings*".

This new approach opened the way to a season of measures aimed at expanding the range of administrative, fiscal and credit incentives and benefits in favour of district enterprises, finally assimilating and equalising all the different types (and also industrial and agricultural districts) in this new notion. It is worth underlining that this is the sole, albeit important, new feature. In moving from the former to the latter definition of district, the government changed political approach, placing more emphasis on the role of entrepreneurial and private initiative than on that of the State. Nonetheless, political discussion about districts changed much more than did the legal approach. In fact the legal definition of production district included – albeit unnecessarily - explicit reference to the two principles of subsidiarity, thereby reconfirming the reciprocity as well as the cooperation necessary between private and public initiatives. In other words, the introduction of this new definition was justified more by general political aims than by a new economic policy stance.

Moving forward in the wake of the political approach above, the third period - in 2008 - saw the first regulation of business networks, which were defined as free aggregations of "single production centres" having the network contract as a specific tool. The network is an important element of the wider category of "institution" and is a basic node in the district conceptual framework very significant for the policy approach (Borràs and Tsagdis [2008]), from a scientific point of view. It may therefore be considered a more focused tool for strengthening the formation of the business network, spontaneously created by firms, thanks to face-to-face interaction, or informal deals or private contracts. Instead, in the political approach, this intervention is deemed the right way of meeting the need for a simpler and faster tool than the district, recognising the changing requirements of mature industrial districts that have significantly changed their relationship with their territory and the local dimension, because companies have adopted delocalisation strategies to better compete in the global market. Rather, DAs and networks should be considered different but integrable instruments.

Over time the government has reformulated several different notions of district in its successive acts without repealing the preceding ones. The Regions are thus able to choose from these notions or mix them according to their own aims and approaches and nowadays they have become a tangled stratification of multiple definitions.

In fact, each Region has regulated the DAs differently (Toccaceli [2012]: 58), often creating a path towards the formula deemed best for its objectives through subsequent interventions designed to respond to multiple pressures - either the changing economic and political scene, or precise requests from the territories concerned to create a specific district or in order to respond swiftly to the flurry of initiatives from the government. For example, following the 2006 Finance Law, ten Regions have enacted laws on DAs, either introducing them from scratch or adapting existing rules to create the conditions for access to new benefits introduced, so as to ensure maximum support to businesses in the Region. Moreover the Regions, far beyond the simple "implementation" of national provisions, have generated a lot of new types of district, e.g. the "supply chain districts" introduced by Lombardy.

Consequently, the overall picture is a heterogeneous image, changing over time, which well reflects the constant and fervent work of the Regions, unceasing and still ongoing, so it is not easy to order such a miscellany.

Therefore, in order to compare regional patterns, it was first of all necessary to identify a truly common concept of DA, highlighting the economic mechanisms actually involved, but taking a different approach than the political and juridical ones.

METHOD

Reconstructing the conceptual framework from a historical, economic and legal standpoint was essential in order to correctly fix the baseline upon which to found a survey on a national scale.

The scope of the survey was limited to the DAs set up in those Regions with a law in force and desk research focused only on the DAs recognised by them.

The methodology is articulated in two phases. The first phase deals with the desk research into the DAs to define their main common features on a national scale, to show how the basic economic mechanisms of the notion of 'district' were designed by the Regions. The second phase deals with an analysis of the possible integrated role of the DAs in the coming programming period.

The desk research is sub-articled, referring to the different objectives: on one hand, to create an updated map of the recognised DAs, on the other hand, to carry out a critical analysis of the complex of regional laws to clarify the very notions of 'district' used by Regions. The latter objective aims to better highlight how the basic economic mechanisms are shaped by the regions and whether their functioning is effectively ensured. Final considerations on these outcomes guided us to the methodological approach, thanks to which a common notion of DA is proposed.

As information about DAs in Italy is not collected, nor immediately accessible, the data used to build Figure 1 came from the official regional sources of information and from examination of the main bibliography on the subject.

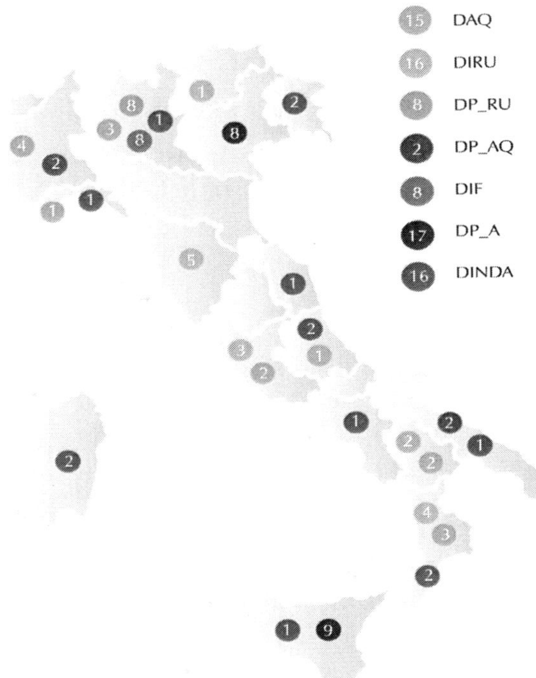

Figure 1: The DAs recognised by the Regions by typology (Source: own elaboration on Regions' Official Journal and official web sites, updated to April 2013).

All regional regulation since 1990 was collated and critically examined in order to learn the essential elements for developing a scheme usable for conducting comparative analysis of such complex regulatory frameworks (Toccaceli [2012]: Appendix). Following this, we were able to highlight the few key elements of each law: the juridical definition of district elaborated or used (derived from those offered by the national laws) and the typologies covered; the purposes attributed to the law; the requirements for identifying DAs and the procedures for their recognition; the organisational models developed to regulate the relations between the different actors involved in districtual dynamics; how projects are planned and financed; how the interventions are monitored.

Critical reflection has suggested considering the juridical issue of clarifying and reducing to a strict number the possible DA typologies, separately from the economic question.

In order to illustrate the connection between the basic economic mechanisms of the district notion and how the Regions have shaped and implemented this mechanism, the overall conceptual framework has been split into its basic components. These are the spatial and value chain dimension, which together represent the economic engine and network making and governance, which together represent the socio-economic engine. All components, of course, work together producing external economies and business environment and generating growth both for businesses and for the Regions where they are located. Finally, there is the political dimension. In this analysis, any external economies generated are the outcome –and even the aim - of a good regulation, rather than an element directly susceptible of regulation.

The spatial and the value chain dimensions in regional legislation are shaped through the choice of how to identify DAs, which relate to localisation and concentration of activities, as well as to their specialisation and interconnection in an organised value chain.

The manifold nature of institutions involved in the district process, as well their relationships, are again outlined through the governance requirements for identifying DAs, as well as in the rules that prefigure a large number of organisation patterns. It is worth underlining that mostly the local network is regulated, perhaps intending a too 'localistic' conception of the district, as better analysed below.

The political component, albeit often deemed external to the conceptual economic framework, in this kind of design is considered essential, because it is by law that life in the districts is regulated. So this component is considered in the sense of policy to be implemented on a local scale and is expressed through the norms for regulating what kind of DA project may be accepted and how they may be supported by public finance. From this analysis derived our methodological approach and the 'methodological' common definition of DAs, which is the milestone for some general reflections about the possible future use of DAs.

In the second phase, this more general concept of DAs is compared with a critical survey of the structural policies for 2014–2020, having regard to the extension of LEADER methodology to all European and Structural Investment (ESI) Funds[k.], as well as to the increased attention paid by rural development policy to cooperation among farms, including in the form of networks and clusters.

[k]'ESI (European Structural and Investment) Funds' are the European Regional Development Fund (ERDF), the European Social Fund (ESF), the Cohesion Fund, the European Agricultural Fund for Rural Development (EAFRD) and the European Maritime and Fisheries Fund (EMFF), which operate under a common framework and work together in a coordinated way according to common provision established by Reg. (EU) 1303/2013.

RESULTS AND DISCUSSION

Findings from Desk Research into the DAs

From the map to the rationalisation types of DAs created by the Regions

Under the legislation discussed above, over time sixteen Regions have recognised 82 districts[l].(Figure 1) in accordance with the restricted number of typologies examined below (Table 2). The recognised DAs are either specialised, hence focused on a specific agricultural or agro-industrial production often with a designation of origin, or non-specialised (Table 3). For reasons given above, there is not a univocal correspondence between typologies of districts and typology of specialisations. For example, some milk and dairy product districts have been recognised as quality agro-food districts, others as districts of the supply chain, others as production or industrial districts. More than half of the DAs (44) specialise in the following sectors: fruit and vegetables (9) plant nurseries (7), animal husbandry (6) wine (5), milk and dairy products (5), cereals, (3) fish (2), agro-energy (2) food biotechnology (2) cork (1) coffee (1) and confectionery (1).

[l]Updated to April 2013.

Table 4: Regional criteria for identifying and recognising DAs (source: own elaboration in Toccaceli[2012])

Indicators	LPS			DP
	DINDA	DAQ	DIRU	All types
Spatial (concentration)	Statistical indicators less restrictive than those for Industrial district		Homogeneous rural identity of the territory	Statistical indicators
	Territorial contiguity	Territorial contiguity	Territorial contiguity	Wider areas also without contiguity
Business (specialization)	Specialised and integrated supply chains	Quality of agro-food production		Economic importance of the sector
		Integrated supply chains	Multi-sector integration	Integrated supply chains
		Territorial integration	Territorial integration	
		Innovation	-	Innovation
	Governance	Governance	Governance	Governance

Toccaceli

Toccaceli *Agricultural and Food Economics* 2015 3:1 doi:10.1186/s40100-014-0019-9

Table 5: DA governance patterns: the bodies and their functions - Own elaboration of regional laws in force

Patterns	Regions	Governance functions	Management functions	Representation	Legal form
Dual	**Veneto (DP)**	Every subject involved in the district (informal network)	Pact Representative		Representation mandate
		Regional Board of Districts and meta-districts	---		---
	Sicilia (DP)	---	Pact Representative		Individual form
		Regional Board of Districts	---		--
	Piemonte (DAQ)	District Desk	Company of District		Company form or associative
	Calabria (DRAQ)	District committee	Company of District		Company form
	Lazio (DRAQ)	District Committee	Company of District		Joint-stock company also consortium
	A.P. of Trento (DIRU)	- Valley Community -Committee of participation	Agency	---	Not required
Single	**Friuli VG (DINDA)**	ASDI			Consortium company
	Abruzzo (DAQ)	Company of District			Capital company
	Basilicata (DRAQ)	District Committee		---	Not required
	Liguria (DAQ)	District Committee		District President	Not required
	Puglia (DP)	District Committee		District President	Not required
Company	**Lombardia (DP)**	---	Company of District		Company form
Network	**Toscana (DIRU)**	Coordinator	---	---	Not required

Toccaceli

Toccaceli *Agricultural and Food Economics* 2015 3:1 doi:10.1186/s40100-014-0019-9

Table 6: A comparison between DA project elements and CLLD strategy elements – own elaboration

Elements	DAs	CLLD (Reg. UE 1303/2013)
Analysis	Ex-ante analysis of situation in district (with help of SWOT analysis), socio-economic situation and situation of main productive sectors.	(a) definition of area and population covered by strategy;
		(b) analysis of development needs and potential of area, including an analysis of strengths, weaknesses, opportunities and threats;
Strategy	Definition of strategic objectives and targets. Identification of main lines of action and innovative actions. Indication of expected results.	(c) description of strategy and its objectives, description of integrated and innovative features of strategy and hierarchy of objectives, including measurable targets for outputs or results. In relation to results, targets may be expressed in quantitative or qualitative terms. The strategy shall be consistent with relevant programmes of all ESI Funds concerned;
Partnership	Identification of public and private stakeholders involved in project and indication of mode of their involvement (usually required to form a legal entity)	(d) description of community involvement process in development of strategy;
Planning	Presentation of single initiatives proposed, sometimes including initiatives of public nature, such as infrastructure of service to the district.	(e) action plan demonstrating how objectives are translated into actions;
	Indication of feasibility of initiatives and realisation times	
Monitoring	Usually an annual report is requested.	(f) description of management and monitoring arrangements of strategy, demonstrating capacity of local action group to implement strategy and description of specific arrangements for evaluation;
Financial plan	Development of financial plan detailing share of contribution required from Region (where it provides direct financing with regional resources) and financial commitments made by each participant and sources of public (EU, national and regional) funding identified.	(g) financial plan for strategy, including planned allocation from each ESI Fund concerned.

Toccaceli *Agricultural and Food Economics* 2015 3:1 doi:10.1186/s40100-014-0019-9

The other non-specialist DAs (38) focus (23) on a *basket of integrated products and services of the territory* (Pecqueur [2001]; Pacciani [2003]; Belletti and Marescotti [2004]) because of the awareness that the district -especially the rural district - determines an overall enhancement of the territory (Iacoponi [2002]; Pacciani [2003]). Typical products with designation of origin (Arfini et al. [2010]) are usually at the core of that basket. This classification does not coincide with that (9) named *multi-product,* which includes those agro-industrial districts less concentrated on integrated action of territorial development, even if focused on several typical products. Among the non-specialised you see also some particular uses of DAs: reinforcing the rural economy of peri-urban areas and getting better governance for managing their relationship with the city (e.g. the rural district of Milan and the other three around the Milan metropolitan area); enhancing the environment and focusing on forest heritage in the Pistoia Mountains in Tuscany. Non-specialised DAs are mostly rural districts, signifying correct use of this tool.

Hitherto, the Regions have recognised a huge number of designations and types of DA for very different economic contexts and rural environments. Unfortunately it is not always understandable from regional laws or DA designations to which kind of juridical notion they refer. So a deeper analysis is needed for correctly re-organising the types of DAs created by the Regions, finding a first criterion to make the multifaceted universe of agricultural districts more homogeneous. The more logical choice seemed to combine the two criteria derived from legal bases: on the one hand, the basic subsequent notions of local productive systems (LPS) and production districts (DP) to which regional laws refer; on the other hand, the basic types of agricultural or industrial district. So all the district types named by the Regions (summarised in Table 4) are reduced to basic typologies on which we can now focus.

Moreover, it should also be noted how the methodological link between LEADER and rural districts (Belletti and Marescotti [2007]) found expression in real life. Because LEADER-eligible areas have the same general characteristic as those where rural districts are recognisable, sometimes some LAGs planned to evolve into rural districts and did so. This happened in two ways, either achieving recognition from their Region (e.g. the LAG-rural district Serre Calabresi or the LAG Golem in Lombardy which led to recognition of DIRU Franciacorta-Sebino-Valtrompia); or, where there were no regional laws, defining themselves or their planning target as a "rural district". Examples of this are the LAG "L'altraromagna" which in 2007/13 developed a local action plan for "the rural district of biodiversity", or several LAGs in the Sardinia Region[m]. Another example comes from the Marche Region where the LAG Colli Esini designed the Rural District Colli Esini as an innovative project to promote its territory. Reading their local action planning documents, the idea emerges that the DIRU are deemed able to boost the image and attractiveness of the area and re-create conditions of economic value creation in a more integrated and stronger way. In some other cases District and LAG paths are distinct but synergistically intertwined (e.g. the Quality Agro-food district of Castelli Romani and Monti Prenestini).

[m]E.g. The LAG "Rural district Barbagia-Mandrolisai-gennargentu-Supramonte" titled its local development plan "toward the rural district" and the LAG "Shardana Lands" titled "Path toward a sustainable rural district and for quality of life".

The DAs' economic mechanisms shaped by the Regions

The Spatial and Business Dimensions

Bearing in mind the evolutionary path along which the legal notion of DAs has evolved over the years, modifying its relation to the scientific concept, this analysis shows how the economic function varies for the different typologies.

The Regions have shaped these dimensions using a long list of criteria to identify and legally recognise the DAs (Table 4). We discuss the four essential types of DA identified, in chronological order and distinguishing the spatial indicators of proximity from the business indicators of businesses' specialisation and concentration.

The first definition of industrial district, later discarded, based on the spatial dimension of "local territorial area", hinged on the concept of local labour system based on commuting elaborated by Sforzi ([1987]). It was identifiable thanks to a set of statistical indicators signifying the concentration of enterprises, entrepreneurs and workers in a very specialised sector which was really important for a certain area.

Held to be too restrictive, after 1999, for the spatial identification of local production systems (LPS), and consequently for agro-industrial districts (DINDA), some discretional and qualitative criteria were added, in order to be able to include many more districts that did not correspond to some of the statistical requirements.

For the rural and agro-quality food districts (DRAQ), the Regions elaborated a rich array of criteria which strictly corresponded to the main points of their legal definition. First, they specified two differentiated requirements: for the rural districts (DIRU), a homogeneous rural identity of the territory as well as integration between agricultural and other local activities; for the quality agro-food districts (DAQ), the presence of quality or typical agro-food products with certified origin as well as interconnection among farms and agro-food enterprises. They also specified two common requirements: the integration of the district's enterprises with the territory, expressed by the production of typical goods, representative of the territory and a well-organised mechanism of governance requested by each Region. Again, there are some links between LEADER and DAs. For example, the Piedmont region has defined its requirements for rural districts as above, but said that they are basically similar to the LEADER programme and cover the same rural zones, so DAs are not recognised in that region to avoid ineffective overlapping.

Like industrial districts, production districts (DP), to be recognised, must demonstrate by a set of statistical indicators their economic relevance and vertical integration of the supply chains. Also the capacity for innovation is required.

Without exception, all the Regions require evidence of governance requirements, that is, a governance process and an effective partnership agreement in place.

Institutions, Organisation and Governance Patterns (Socioeconomic Dimension)

The Regions have taken different political approaches (not part of this analysis, as mentioned above) in dealing with the wide category of institutions which may be public, semi-public and private actors, also non-local, and their possible networks, which may be formal, semi-formal or informal. The Regions have produced a broad range of possibilities. To cite an example, Tuscany has a network approach, and its regulation is very confident about the local network's effectiveness, so there is little legislation about governance. Conversely, a greater group of Regions that have an implicit top-down approach have much more regulated and formalised partnership arrangements. This group includes those Regions that have shaped DAs as effectively implementing rural development policy, also conferring upon them funding.

Apart from this rough distinction, it is very hard to recognise any more detailed pattern by which to organise the district's bodies. More or less formalised, those bodies are conceived for working on two basic functions: governance at local level, to identify aims and draw up the project; and management, to implement the district project and strategy. Considering these basic functions only, we can distinguish (Table 5) either a *single pattern* for those Regions which envisage only one body (usually named District Committee) to carry out both functions for the district (in some cases the president of the district is the representative); or a *dual pattern,* when two distinct bodies are planned, one for each function. Under the 2006 Finance Law, some Regions have requested a Representative of the district to be identified, charged with interfacing with the public administration at various levels. In the dual pattern, this may be the person charged with management functions, in the single pattern, when it occurs, it may be the district committee or an addition to it. A different pattern is followed by Lombardy whose DAs are composed entirely of enterprises, so the DAs are organised just by a District Company. A legal form to constitute those bodies is established in some Regions with various more specific requirements.

The very articulated, differentiated and complex procedures for promoting, constituting and sometimes formalising the governance process and partnership-forming often require proof of the bottom-up character of this process, even when an implicit desire for control is easily detected. This demonstrates a shared acknowledgement of the importance of trusting in involvement of local and private actors for a successful district.

The Policy Implementation Dimension

Almost paradoxically, at least in relation to the conceptual framework, the Veneto and Sicilia Regions consider the development of the district project one of the most important requirements for recognising a district. For them, the district practically coincides with or is embedded in its project, as confirmed by the unique procedures for financing districts established by Veneto, described below. It is just an example, but one that says something about the conception and the concrete expectations about this policy tool.

This brings us into the operational dimension, as the project (also called pact or plan or programme) always has clearly expressed rules, for achieving the real aim of mobilising local resources towards innovative actions along medium-term pathways. In fact "district project" invariably means a group of individual or collective projects, always private enterprises, sometimes public too, together coherent with the main objective of the strategy. Usually the project has to be 'ready to be implemented' so it has to include feasibility studies and, when specific finance is provided, an economic-financial plan. Some regional regulations require projects to be subscribed by all the participants in the district, to assure their commitment to working together, to investing and (where needed) to co-financing as agreed. In those cases the project becomes the object of a binding contract.

Even if the procedures for elaborating the project are different, its elements (Table 6) are always the same: a socio-economic analysis of the districtual area, the chosen objectives and the strategy for achieving them, the partnership composition, a plan of action and the financial plan. Monitoring, always prescribed, is not much stressed. The project is evaluated and, if approved, financed by the Region.

So the project has three main features. It is always *integrated,* either in a vertical way in the case of districts mainly focused on products and supply chain, or in a horizontal way in the case of districts mainly focused on local development. It is *strategic*, so in the specified time of three years (the same in all Regions) it focuses only on the chosen strategic goal. It is *programmatic* in all those Regions which finance, directly or not, the districts' projects. In this case the districts are a reference point for regional programming action. Finally, the project has usually (but not always)*bottom-up* promotion, again indicating that generally the Regions recognise that a good project - and its chance of success – requires good dialogue among the actors. As a matter of fact, a number of Regions support the preparatory actions for communicating the aims of the district and involving most of the actors.

It is observed that the more or less implicit desire for control by the Regions over the district's actions is strictly related to the acknowledgement of a specific line of credit to the DA project. The case of Veneto is emblematic, which provides a specific three-year credit line, so that the DA's actions, realisations and financial reports are always under regional control.

For financing, too, it is very hard to identify any pattern (Table 7), as every Region has followed its own path regarding the procedures, juridical tools and the aspect to be supported.

Table 7: Regional choices regarding financing DA projects – own elaboration based on regional laws in force

Regional support only to preparatory actions	Negotiated planning tools	Financial line of the regional budget	Reserved share of regional RDP
	Friuli V.G.	Friuli V.G.	
		Veneto	
Lombardia	Lombardia		
	Puglia		Puglia
Sicilia		Sicilia	
	A.P. Trento	P.A. Trento	
Lazio	Lazio	Lazio	
	Abruzzo		Abruzzo
Liguria			
Piemonte			
	Calabria	Calabria	
Basilicata	Basilicata		Basilicata
Toscana	Toscana		

Toccaceli

Toccaceli *Agricultural and Food Economics* 2015 3:1 doi:10.1186/s40100-014-0019-9

Regarding financing of projects, there are three specific ways. One is to use negotiated planning tools. In other cases, a few Regions provide a specific line of credit within the regional budget, so that recognising the district, approving the project and granting funding may be contextual. Some other Regions reserve a share of the regional rural development plan for DAs.

Sometimes DA projects are included in regional planning as specific actions to implement one or more policies, in an integrated way and following multiple aims (e.g. rural development and environmental sustainability or enhancing firms' competitiveness and internationalisation and innovation processes).

In all events, most Regions consider coordinated use of existing policies to implement the district project as the culmination of the district's path.

The Methodological Approach to DAs

Having completed the desk research, we can summarise the main findings and attempt to identify some logical consequences.

It may be generally stated that the Regions see DAs as supporting the competitiveness of agricultural enterprises and their innovation processes, strengthening the agricultural and agro-food supply chains and opening them to international markets, developing rural areas, protecting the environment and managing the relationship between city and countryside in peri-urban areas.

More specifically it may be observed that those aims are pursued in a differentiated way, so that there is a range of variation in the understanding of DAs' functions and applications. These range from the totally horizontal, conceived for aims of rural development and environmental sustainability, to the totally vertical focused almost exclusively on the competitiveness of farms and firms in the agricultural and agro-food supply chains and their innovation processes. It is self-evident that this presumed assignation makes more political than scientific sense. This misconception may have thus influenced the correct use of this policy tool.

We have reduced the array of DA designations made by the Regions to four categories derived from the main legal basis, in order to obtain an initial criterion to proceed with our analysis. Although useful, this criterion was not enough to read such a complex phenomenon, e.g. for univocally linking any typology to a function or a way of financing DAs.

Without exception, the Regions have established such a set of criteria as to allow the development of the economic process, selecting districts with requirements of proximity, concentration and specialisation, coherently differentiating them in function of their main orientation towards the horizontal or the vertical dimension.

We note a variety of approaches to network, ranging from the informal involvement of multiple local actors. At the other extreme, the district coincides with the local partnership agreement, despite the fact that the operational solutions adopted for recognising and making operative the network are generally very complex and time-consuming. This variety of approaches was our primary reason for attempting a methodological approach for finding a common basic definition of Italian DAs.

This impression is reinforced by the fact that a prerequisite for the recognition of DAs and the basis of their successful functioning in all regional laws is the implementation of broad governance processes, mostly at local level.

To concretise the strategy, the highest aim of the DAs is developing and implementing a medium-term project that is strategic and integrated. In most cases, it is also programmatic, in the sense that, in several Regions, DA projects become part of the process of regional planning policy, so they are also considered as a source of input information for the purposes of policy planning, management and land use planning.

Finally, we note that all the Regions introduced monitoring of the application of their own laws, although specific indicators of effectiveness are almost never mentioned. As a result no data - no comparable data - are available to enable objective analysis and deduction.

This homogeneous method adopted by the Regions, despite differences and changes of political vision and leadership, was possible because, as has been observed (Toccaceli [2012]), the methodological approach defined by the law on industrial districts has remained at the heart of all subsequent national legislation on DAs. Considering the subsequent assimilation of all kinds of district, it has been pointed out that the government has implicitly followed a methodological framework which remained constant and neutral in regard to various economic sectors of application and economic objectives, for which districts were prepared and used over time.

As a consequence, a methodological definition (see Figure 2) is proposed that qualifies DAs as a method of intervention at local level, based on a multi-level governance dynamic (Commission of the European Communities [2001]; Committee of the Regions [2009]). This comprises the local partnership – established among districtual firms and private and public local stakeholders and operating by a participatory method – which formulates a strategy and develops innovative projects for enhancing the districtual productive system. There is also a district representative, to dialogue with the Public Administration at various levels, even if informally identified among the local actors of the district. Through a specific juridical tool (in the form of a contract, e.g. the district pact) the projects are included in higher-level programming tools, in application of the principles of vertical and horizontal subsidiarity, so they are financed with public and private sources and implemented. This definition focuses on the complex system of relationships linking the different actors at local and also at higher level, where the privileged interlocutor is the Region, through which policies are passing as a result of the national devolution process and application of the principle of horizontal and vertical subsidiarity.

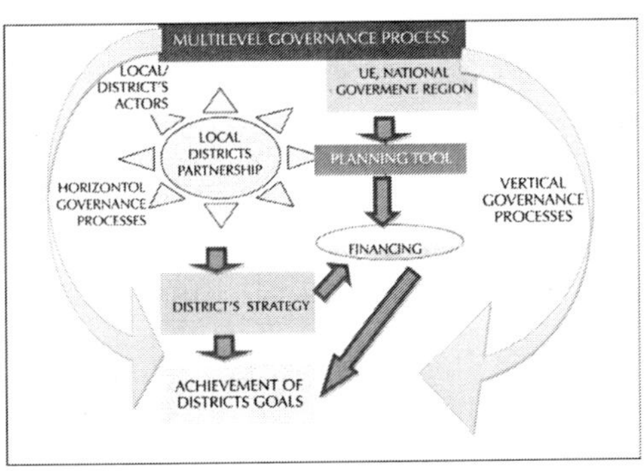

Figure 2: The general methodological scheme of DAs (Source: Toccaceli D. 2012:46).

The district thus defined, because of the principles that animate it, is consistent with post-2013 rural development policy, both for its participatory approach and for the local scale of intervention which is significant to implement European Structural and Investment Funds, more so than in the past.

The DAs in the Perspective of the 2020 European Goals for Agriculture and Rural Development

At the time of Agenda 2000, rural and quality agro-food districts were established within the context of a significant set of tools to modernise and strengthen Italian agriculture, in the face of new scenarios of increased exposure of firms to global competition. Two programming cycles later, Italian agriculture and rural territories still need to be strengthened, including through new organisational solutions (Canali [2013]:79) to enhance their competitiveness.

More in general, in the period of Europe 2020 strategy, old and new challenges have to be faced by means of the CAP and the rural policy, as well as the other ESI Funds (De Castro [2010]; De Castro et al. [2011]; De Filippis [2012]). These include food security and food quality concerns generating great demands on food systems, environmental concerns linked to the loss of biodiversity, landscape deterioration and the demand for renewable sources of energy.

The EU has reinforced its strategic approach, linking all ESI funds under common provisions and a set of guiding principles, the Common Strategic Framework (CSF), as well as establishing eleven Thematic Objectives on which ESI Funds support should be concentrated. The strategy is based on (among others) the principle of multilevel governance and partnership. So, in utilising ESI Funds, every Member State is requested to prepare a Partnership Agreement reached with the Commission and a broad national partnership.

In the light of this new orientation of European policies and considering the comprehensive methodological definition of DAs, we can attempt to give a detailed answer to our main research question. We asked ourselves whether and how DAs may be still useful in new programming policies, with reference to structural policies, more in particular to rural development policy and with special attention to the relation between DAs and LEADER method and programmes.

More so than in the past, the local level is key in implementing policy measures. Indeed a specific chapter of the CSF Regulation is dedicated to community-led local development[n.] (CLLD) which supports local strategies led by a Local Action Group (LAG) and thanks to multi-fund financing. The device established by the EU is modelled on LEADER methodology so, analysing point by point the main features of CLLD, it is not surprising to find great similarity between CLLD methodology and DA principles. The new possibility of carrying out integrated and multi-sectorial area-based development strategies and of supporting them by multi-fund resources is the most significant similarity and a new opportunity for DAs too (especially for Rural Districts). In fact this new tool becomes more important given the different types of rural territory, especially considering the wide-ranging 'ruralities' from remote to peri-urban rural areas. In the new context, there should not be a dichotomy between urban and rural, rather a more proactive relationship. Starting anew from their own endogenous potential, rural areas are called on to develop "smart specialisation" based both on their territorial capital and local actors' capability to act together to put in place successful strategies.

[n]Reg. (EU) 1303/2013 Article 2(19).

Focussing on rural development policy, the European agricultural fund for rural development (EAFRD) contributes to achieving specific objectives[o.] (agriculture competitiveness, environment and rural territories development) following detailed priorities[p.] in accordance with the CAP framework. So this Fund has established precise measures which may be considered useful for DAs, or which may be more effectively implemented where there is a "district atmosphere" and thanks to DAs. Within the EAFRD framework, strong attention is paid to the role played by co-operation[q.], so the related measure, reinforced and enlarged in comparison with the past and intended to stimulate innovation, is very complex and multi-purpose. In fact, the concept of innovation is very broad and the possible forms of co-operation are highly articulated with respect to the possible subjects involved, in order to respond to all the strategic objectives of the fund: environmental, competitive and territorial balance (Table 8). For this paper's purpose, we note that co-operation can involve different actors from the food chain, agriculture or forestry sectors, as well as producer groups, cooperatives and inter-branch organisations. Co-operation may also be intended to "create clusters or networks[r.] ", even if they have to be "newly formed" and "commencing an activity that is new to them", in order to pursue the aim of innovation.

[o]Reg. (UE) n. 1305/2013 Article 4.

[p]Reg. (UE) n. 1305/2013 Article 5

[q]Reg. (UE) 1305/2013, Article 35.

[r]Reg. (UE) n. 1305/2013, Article 2(1,q) «"cluster" means a grouping of independent undertakings, including start-ups, small, medium and large undertakings as well as advisory bodies and/or research organisations - designed to stimulate economic/ innovative activity by promoting intensive interactions, the sharing of facilities and the exchange of knowledge and expertise, as well as contributing effectively to knowledge transfer, networking and information dissemination among the undertakings in the cluster».

Table 8: The 'bundle' measure of co-operation in EAFRD, Reg. (UE) 1305/2013 Article 35 (2), referred to EAFRD general objectives Article 4 – (source: Toccaceli[2012]:101)

Eafrd objectives article 4	Co-operation actions supported By Eafrd Article 35
(a) fostering the competitiveness of agriculture	(a) pilot projects;
	(b) the development of new products, practices, processes and technologies in the agriculture, food and forestry sectors;
	(c) co-operation among small operators in organising joint work processes and sharing facilities and resources and for the development and/or marketing of tourism services relating to rural tourism;
	(d) horizontal and vertical co-operation among supply chain actors for the establishment and the development of short supply chains and local markets;
	(e) promotion activities in a local context relating to the development of short supply chains and local markets;

(b) ensuring the sustainable management of natural resources, and climate action	(f) joint action undertaken with a view to mitigating or adapting to climate change;
	(g) joint approaches to environmental projects and ongoing environmental practices, including efficient water management, the use of renewable energy and the preservation of agricultural landscapes;
	(h) horizontal and vertical co-operation among supply chain actors in the sustainable provision of biomass for use in food and energy production and industrial processes;
	(j) drawing up of forest management plans or equivalent instruments;
(c) achieving a balanced territorial development of rural economies and communities including the creation and maintenance of employment	(i) implementation, in particular by groups of public and private partners other than those defined in point (b) of Article 32(2) of Regulation (EU) No 1303/2013, of local development strategies other than those defined in Article 2(19) of Regulation (EU) 1303/2013 addressing one or more of the Union priorities for rural development;
	(k) diversification of farming activities into activities concerning health care, social integration, community-supported agriculture and education about the environment and food.

Toccaceli

Toccaceli *Agricultural and Food Economics* 2015 3:1 doi:10.1186/s40100-014-0019-9

Support may also be given to the "implementation, in particular by groups of public and private partners of local development strategies"[s] adding it to the CLLD and limiting it to EAFRD support. So it seems that many types of DA and DA project might be covered by this measure, if Regional planning allows and provides for this.

[s]Reg. (UE) n. 1305/2013, Article 35, (2,i).

Furthermore, a broad typology of cost is eligible and "where a business plan or an environmental plan or a forest management plan or equivalent or a development strategy is implemented, Member States may grant the aid either as a global amount covering the costs of co-operation and the costs of the projects implemented or cover only the costs of the co-operation and use funds from other measures or other Union Funds for project implementation"[t], obviously avoiding any over-compensation. It should be underlined that many different contexts and purposes can be adapted into the bundle of this detailed and flexible measure. Also, DAs might strengthen their partnership or renew their project for innovating, whether they are quality agro-food, or rural, or aiming to manage relationships between rural and urban areas, or focused only on a single product or food supply chain, considering that these tools depend on the good functioning of multi-level governance. Generally speaking, it might be more effective in those cases where governance pathways are already set out.

[t]Reg. (UE) n. 1305/2013, Article 35, (6).

Finally, the multiple contact points between DAs and LEADER suggest two different points.

On one hand, "district thinking" may suggest some indications about LEADER Programmes newly financed by EAFRD. The European Court of Auditors ([2010]), among other more specific observations, pointed out that local action groups' (LAGs) actions were not well focused on local strategic objectives, and that the added value due to the LEADER approach was not seen to be achieved. Even if those recommendations have been taken into account by the European Commission and Member States to formulate better rules to ensure those gaps will be filled, in practice, "district thinking" may be useful in better concentrating local strategies, and resources as well, on the most active value chains that may activate interactions among enterprises of all sectors or belonging to a same value chain.

On the other hand, "LEADER thinking" may help to point out some strong and weak points in DAs. Our research shows that DAs shaped by Regions are always conceived as competitive development tools to enhance agricultural and agro-food assets and to reinforce the local rural system, Thanks to our new-found methodological approach, we can say that DAs, by design, have proved to be:

A flexible method. Based on different and changing political inspirations, Regions have adjusted the district tool to fit many different cases, given the high degree of autonomy that now exists following extension of the LEADER method to ESI Funds.

A homogeneous method. Though each Region was following its own purposes, together they have inspired and contributed to formulating a methodological pattern with common and homogeneous features.

A stable structure for organising the local economy. The fact that all Regions ask DAs to make updatable three-year projects indicates that DAs are designed as a stable structure designed, in many cases, to become a constant reference point in the programming of regional policies. It is true that the complex mechanisms required to set up DAs may be so tiresome as to envisage the possibility of numerous failures or dropouts. However, the very amount of time spent in setting up DAs may be considered a strong point if it means practising dialogue among local stakeholders. So DAs can be regarded as a flexible device for policy delivery.

Conversely, the most evident weak point of DAs is that, more than ten years after the Law of Orientation, it is still impossible to know more about the DA phenomenon because there is no constantly updated information about regional choices, regional legal sources or DAs actually operative, nor is there a regional system for monitoring DA performance.

In comparison with the LEADER method one may evaluate DAs' critical points, looking at three "genetic" weaknesses:

DAs do not form a network among themselves. So, within their own regulation, they have neither territorial cooperation nor a place to exchange their experiences.

Only some DAs can rely on a certain perspective of financial support, as successful implementation of medium term planning is requested for this.

They lack a monitoring system by third parties. Every Region has made monitoring compulsory by law. But - especially in cases where the law already provides for a regional non-financial line dedicated to the DA - there are no actual evaluations of effectiveness of the method, nor even homogeneous surveys able to present more generalised elements of assessment.

So, even if still an expanding phenomenon, the DA method lacks an evaluation system, such as that of LEADER, to permit learning from mistakes and the updating itself, rather than fostering ineffective consumerism of policy tools.

CONCLUSIONS

This article aimed to create the conditions for carrying out research on national scale into DAs, starting from an unclear political, juridical and economic context. Thanks to the "reconstructive reading" of the legal bases as well as of the conceptual economic framework, and to desk research into DAs shaped by regional laws, we analysed how the basic components of the district conceptual framework were elaborated by the different Regions in such a way as to ensure the conditions for displaying the potential of the 'district process'. Even if we could not identify a specific pattern because of the different political approaches developed over time, a methodological approach was found that allowed us to outline the DAs' common points and their economic mechanisms and to collect them into a general definition.

On this basis, it was possible to develop a general reflection on the possible future use and utility of DAs, integrating them into the coming programming period. At the same time, some of these reflections on DAs may offer suggestions for other Countries where the same regulation is in force. So, the chances for the coming years are linked to different scenarios, whether the main purpose is either reinforcing weak and marginal rural systems or enhancing competitiveness of strategic supply chains. Two ways are possible:

assimilating DAs and LAGs indicated for developing and implementing multi-fund local strategies (CLLD). This depends on the success of DAs , i.e. whether DAs emerge as an example of stable organisation of the local economy and good governance or not.

including DAs in the 'bundle' of the co-operation measure : constituting new clusters and networks for the wide range of aims covered by the law; referring to supported implementation by groups of public and private partners of local development strategies.

These new opportunities would allow Regions to overcome some of the weaknesses highlighted and have systems in place for standardised, constant monitoring and evaluating in the medium term. Regions could also make use of the results of evaluation and monitoring, which would be available for the first time; to improve DA methods, just as the LEADER method was improved over time.

In conclusion we remark that this paper refers to the findings of the first survey of DAs, developed on national scale. Although it contributes towards classifying some early points, many questions remain open for further research.

ACKNOWLEDGMENTS

This paper is possible thanks to research promoted by the National Rural Network, coordinated by Mrs. Graziella Romito, within the Task Force "Integrated Planning" 2007-2013. Special thanks are due to Professor Alessandro Pacciani and Professor Ferdinando Albisinni for their precious observations during the research period which has resulted in this paper. I am very grateful to the two anonymous referees for their stimulating considerations. My thanks, too, to Julia Campbell Hamilton for proofreading the paper. Any errors remain, of course, my responsibility alone.

REFERENCES

1. Adornato F (Ed) (2005) Sviluppo locale e contrattazione programmata Franco Angeli Milano.
2. Aimone S., (Ed) (2006) Programmazione integrata e sviluppo rurale del Piemonte in: Quaderni di ricerca Istituto di Ricerche Economico e Sociali del Piemonte (IRES),Torino.
3. Albisinni F (2010) Distretti e sviluppo rurale: elementi per una lettura delle regole di diritto, in: Agriregionieuropa, Anno 6n.20, pp.21-25 ISSN: 1828–5880
4. Albisinni F (2011a) Distretti e contratti di programma in agricoltura in Costato L. Germanò A. Rook Basile E. (Eds) Trattato di diritto agrario, Volume Primo:409–461 UTET, Torino
5. Albisinni F (2011b) Prodotti agro-alimentari e distretti in agricoltura In: Goldoni M, Sirsi E. (eds) Il ruolo del diritto nella valorizzazione e nella promozione dei prodotti agro-alimentari: Giuffrè Milano pp103–148 ISBN: 88-14-17236-6
6. Angeli L (2000) L'atmosfera di Artimino e noi economisti agrari. La Questione Agraria n.3:111–121 ISSN: 1593-8441
7. Arfini F, Belletti G, Marescotti A (2010) Prodotti Tipici e Denominazioni Geografiche: Strumenti di Tutela e di Valorizzazione. Quaderni, Gruppo 2013, Roma.
8. Baldi L., Casati D (2009) Un distretto della quarta gamma? Il comparto che vende tempo libero, in: Agriregionieuropa, Anno5, n. 16, pp. 50-52, ISSN: 1828-5880

9. Becattini G (1987) Mercato e Forze Locali: il Distretto Industriale. Il Mulino, Bologna.
10. Becattini G (1989) Riflessioni sul distretto industriale marshalliano come concetto socio-economico. Stato e mercato 25:111-128
11. Becattini G (2000) "Distrettualità" fra industria e agricoltura in La Questione Agraria n.2, pp: 11–24 ISSN: 1593-8441
12. Becattini G (2001) Distretti e ruralità: sfide al riduzionismo economico. Una replica in: La Questione Agraria n.1pp. 119-127. Franco Angeli, Milano, ISSN: 1593-8441
13. (2009) A Handbook of Industrial Districts. Edward Elgar, Cheltenham, UK – Northampton, MA, USA.
14. Bellandi M (1987) La formulazione originaria in Becattini G., Mercato e forze locali: il distretto industriale: Il Mulino, Bologna
15. Belletti G, Marescotti A (2004) Il Distretto Rurale: Metafora Organizzativa e Strumento di Governance. In: Hoffmann A (ed) Esperienze di Programmazione Dello Sviluppo Locale: il Caso del Parco dei Nebrodi, Franco Angeli, Milano.
16. Belletti G, Marescotti A (2007) Il Distretto Come Strumento di Governance per lo Sviluppo Rurale. In: Rete Leader, Leader e distretti rurali: sinergie e complementarietà. Biemmegraf, Macerata.
17. Belletti G., Marescotti A (2010), Il distretto rurale, In Pacciani A., Toccaceli D. (a cura di) Le frontiere dello sviluppo rurale. L'economia grossetana tra filiere e territorio, Franco Angeli, Milano, ISBN: 978-88-568-3268-6
18. Borràs S, Tsagdis D (2008) Cluster Policies in Europe. Edward Elgar, Cheltenam, Uk – Northampton, MA, USA.
19. Canali G (2013) L'OCM Unica e le Misure di Mercato. In: Filippis F (ed) La Nuova PAC 2014–2020. Un'analisi Delle Proposte Della Commissione, Quaderno, Gruppo 2013, Roma.
20. Caré V, Reda E, D'Angelillo E (2010) L'esperienza dei distretti rurali ed agroalimentari in Calabria. Evoluzione normativa ed iniziative proposte per un nuovo modello di governance e di sviluppo delle aree rurali. Rete Rurale Nazionale, Roma.
21. Casati D (1988) Come cambia l'agricoltura nel sistema agro-industriale in Strategie e adattamento nel sistema agro-industriale INEA, Roma

22. Casati D, Banterle A, Baldi L (1999) Il distretto agro-industriale del riso. Franco Angeli, Milano.
23. Cecchi C (1988) distretto industriale: l'agricoltura dalla complementarità alla dissociazione. La questione agraria n.32, pp 91-123, Franco Angeli, Milano, ISSN: 1593-8441
24. Cecchi C. (1992) Per una definizione di distretto agricolo e distretto agroindustriale La Questione Agraria, n. 46, pp 81–107, Franco Angeli, Milano, ISSN: 1593-8441
25. Cecchi C (2000) "E se facessimo tanti bei campi da golf?": ovvero Becattini e la campagna La Questione Agraria, n.3, pp 123–132, Franco Angeli, Milano, ISSN: 1593-8441
26. Cecchi C, Cianferoni R, Pacciani A (1991) Economia e politica dell'agricoltura e dell'ambiente. CEDAM, Padova.
27. Cianferoni R (1990) L'agricoltura e l'ambiente nel distretto industriale di Prato. Accademia dei Georgofili, Firenze.
28. Cnel (1999) I protagonisti e le esperienze locali di programmazione negoziata. CNEL, Roma
29. CNEL (2007) Distretti rurali ed aroalimentari di di qualità: il manifesto di Matera alla luce delle nuove politiche di sviluppo rurale. Osservazioni e proposte Assemblea CNEL del 20 Marzo, CNEL, Roma
30. Collignon P (2014) Le Syndicate intercommunal pour la promotion du canton de Clerveux "SYCLER" au service du développementglobal d'un territoire rural In: Pacciani A., Toccaceli D. (Eds) Percorsi di governance per la valorizzazione delle aree rurali nella prospettiva di riforma delle politiche europee, pp61-65 I Georgofili Quaderni 2013-II Società Editrice Fiornetina, Firenze ISBN: 978-88-6032-294-4
31. Commission of the European Communities (1988) The future of rural society, COM(88/501) Bruxelles: Bullettin of the European communities, Supplement 4/88
32. Commission of the European Communities (2001), European governance – A white paper COM(2001) 428 def., Bruxelles: Official Journal of the European Communities, C287/1, October, 10

33. Committee of the Regions (2009) The Committee of the Regions'White Paper on Multilevel Governance Own-Initiative Opinion, Bruxelles: 80th plenary session, 17 and 18 June
34. Conti L. (2005) Le politiche distrettuali della Regione Piemonte in: Agriregionieuropa Anno1, n. 2, pp. 39-41, ISSN: 1828 - 5880
35. Contò F, La Sala P (2011) Approccio Territoriale e Sviluppo Locale. Il Programma di Sviluppo del Distretto Agroalimentare di Qualità del Metaponto. Franco Angeli, Milano.
36. Contò F, La Sala P, Papapietro P (2010) Identità, Qualità e Competitività dell'area del Pollino: un Caso di Filiera "Atipica" Agro-Territoriale. Atti della XXXI Conferenza scientifica annuale AISRE, Aosta.
37. Corò G, Grandinetti R (1999) Strategie di delocalizzazione e processi evolutivi nei distretti industriali italiani. L'industria AnnoXX 4:897-924
38. Courlet C., Pecqueur B (1992) Les systèmes industriels localisés en France: un nouveau modèle de développement In: Benko G., Liepitz A. (éds), Les régions qui gagnent. Districts et résaux: les nouveaux paradigmes dela géographie économique, PUF, Paris, pp.81–102
39. De Benedictis M (2000) Economia agraria e distrettualità in La Questione Agraria n.2, Franco Angeli, Milano, ISSN: 1593-8441 pp. 25–34
40. De Castro P (2010) L'agricoltura Europea e le Nuove Sfide Globali. Donzelli Editore, Roma.
41. De Castro P., La Via G., Adinolfi F., Marangon F (2011) Il processo di riforma della Politica Agricola Comunitaria Stato dell'arte e prospettive. In: International Agricultural Policy, n. 1, Edizioni L'informatore Agrario. Verona, ISSN: 1722-4365.
42. De Filippis F (2000) Distrettualità e politica agraria in La Questione Agraria n.2, Franco Angeli, Milano, ISSN: 1593-8441
43. De Filippis F (Ed) (2012) La nuova PAC 2014–2020. Un'analisi delle proposte della Commissione Quaderni, Gruppo 2013, Roma
44. De Meo G., Nardone G (2002) Programmazione negoziata e sviluppo rurale in Basile E., Romano D., Sviluppo rurale: società, territorio, impresa, Franco Angeli, Milano

45. De Rita G., Bonomi A., (Eds)(1998) Manifesto per lo sviluppo locale Bollati Boringhieri, Torini
46. De Rosa M, de Vincenzo D (2002) Stato e Dinamica dei Sistemi Produttivi Territoriali nel Rapporto Locale/Globale. In: Rosa M, Vincenzo D (eds) Tra globalizzazione e localismo, Liguori, Napoli.
47. European Union (2013) Official Journal of the European Union, L 347, Volume 56, 20 december 2013. ISSN: 1977–0677
48. European Commission's Directorates General AGRI, EMPL, MARE and REGIO (2013) Common Guidance on Community-led Local Development in European Structural and Investment Funds., http://ec.europa.eu/regional_policy/informing/dialog/pdf/clld_guidance_2013_04_29.pdf (last access 15/04/2014)
49. European Court of Auditors (2010) Implementation of the Leader approach for Rural Development Luxembourg: Publication Office ISBN: 978-92-9207-883-6
50. Fabiani G (2000) Distretti o Sistemi Agricoli Locali? In: La Questione Agraria. Franco Angeli, Milano. pp 33-41 ISSN: 1593-8441
51. Fanfani R, Montresor E (1998) Istituzioni e imprese nel percorso di sviluppo dei sistemi locali di produzione agroalimentare. La questione agraria n.69 pp.87-108 Franco Angeli, Milano ISSN: 1593-8441
52. Favia F (1992) L'Agricoltura nei sistemi produttivi territoriali. La questione agraria n.46 pp.109-136 Franco Angeli, Milano ISSN: 1593-8441
53. Favia F. (2000) Economia agraria e distrettualità: variazioni sul tema La Questione Agraria, n.3, pp.133-145 Franco Angeli, Milano, ISSN: 1593-8441
54. (2008) Il distretto della pesca di Mazara del Vallo. Una buona pratica di cooperazione tra aziende internazionali. Franco Angeli, Milano.
55. Feser E., Isserman A (2009) The Rural Role in National Value Chains In: Regional Studies, 2009, vol. 43, issue 1, pp. 89-109, doi:10.1080/00343400801968429
56. Forleo M (2000) I Sistemi Agricoli Territoriali. In: CNEL, Secondo rapporto sull'agricoltura. L'agricoltura tra Locale e Globale.

Distretti e Filiere vol N.23.Collana Documenti. CNEL, Roma. pp 295-324

57. Gallego Moreno F.J. (2014) La estructuraciòn y dinamizaciòn social para una mejor goberannza de las comunidades rurales: Caso de studio en Cuenca, Espana In: Pacciani A., Toccaceli D. (Eds) Percorsi di governance per la valorizzazione delle aree rurali nella prospettiva di riforma delle politiche europee,pp29-49, I Georgofili Quaderni 2013-II Società Editrice Fiornetina, Firenze ISBN: 978-88-6032-294-4
58. Garofoli G (2011) La "Governance" Dello Sviluppo Locale Nelle Langhe, in: Seminario Nazionale su Politiche, Istituzioni e Sviluppo Rurale:Come Migliorare i Processi di Governance. Rete Rurale Nazionale, Roma.
59. Gulisano G, Marcianò C (2008) I distretti rurali in Calabria Università degli Studi Mediterranea di Regio. Dipartimento di Scienze e tecnologie Agro-forestali e Ambientali, Calabria.
60. Hoffmann A (2008) I distretti rurali in Sicilia Università Di Palermo, Dipartimento di Economia dei Sistemi Agro-forestali. Fotograf, Palermo.
61. Iacoponi L (1990) Distretto industriale marshalliano e forme di organizzazione delle imprese in agricoltura in Rivista di Economia Agraria Anno XLV, n.4 pp 712-734 n. 4 , Franco Angeli, Milano. ISSN 0035-6190
62. Iacoponi L (2001) Distrettualità agricola: una difficile (e breve) navigazione tra opposti paradigmi economici in La Questione Agraria n. 4:111–118, Franco Angeli, Milano, ISSN: 1593-8441
63. Iacoponi L (2002) Dal distretto agricolo al distretto rurale in Valorosi F. (a cura di), Lo sviluppo del sistema agricolo nell'economia post-industriale: Franco Angeli, Milano
64. Lee J., Arnason A., Nightingale A., Shucksmith M (2005) Networking: social capital and identities in European rural development. Sociologia ruralis, 45 (4):269-283 ISSN: 1467-9523
65. Maizza A (2010) I Distretti Agroalimentari nel Contesto Globale. Un'analisi sul Territorio Pugliese. Franco Angeli, Milano.
66. Marangon F (2008a) Aspetti economici per l'individuazione di un Distretto Rurale nel territorio della Pianura Friulana, CCIAA

di Udine, Convegno Ipotesi di un distretto per lo sviluppo rurale della Bassa Pianura friulana, Agriest Udine, 28 gennaio

67. Marangon F (2008) Lo sviluppo rurale sostenibile in Friuli Venezia Giulia: tra distretti rurali e distretti di economia solidale, Seminario L'economia locale per lo sviluppo del territorio. Università degli Studi di Scienze Gastronomiche, Pollenzo (Cn).
68. Mollard A (2001) Qualité et développement territorial: une grille d'analyse théorique à partir de la rente. In: Économie rurale. N°263, 2001. pp. 16–34. DOI: 10.3406/ecoru.2001.5240,http://www.persee.fr/web/revues/home/prescript/issue/ecoru_0013-0559_2001_num_263_1
69. Montresor E (2000) I sistemi locali di produzione agroalimentare CNEL, Secondo rapporto sull'agricoltura. L'agricoltura tra locale e globale. Distretti e filiere. Collana Documenti CNEL, II serie. N.23,Roma
70. Montresor E., Pecci F., Pontarollo N. (2011) Distretti agroalimentari di qualità e processi di governance locale: due regioni a confronto in Agriregionieuropa, Anno 7, n. 24 pp 57-61. ISSN: 1828 - 5880
71. Murdoch J (2000) Networks – a new Paradigm for Rural Development? J Rural Studies 16(4):407-419 http://dx.doi.org/10.1016/S0743-0167(00)00022-X
72. Musotti F (2001a) Il territorio: da sempre nell'analisi economico-agraria italiana La Questione Agraria, n. 4:119–131, Franco Angeli, Milano, ISSN: 1593-8441
73. Musotti F (2001b) Il "nuovo" orientamento dell'agricoltura italiana: riscoperta del territorio, ruralità, distrettualizzazioni Sviluppo locale, VII, 17:79–91 , Rosenberg & Sellier, Torino. ISSN:1974-2193
74. Oxford Research AS (2008) Cluster policy in Europe Oxford research AS, January http://www.clusterobservatory.eu/system/modules/com.gridnine.opencms.modules.eco/providers/getpdf.jsp?uid=100146
75. Pacciani A (1997) Il progetto "Maremma Distretto Rurale d'Europa" Atti dell'Accademia dei Georgofili, Anno 1997. Industrie Grafiche Fiorentine, Firenze.

76. Pacciani A (2002) Il Distretto rurale della Maremma: dalla proposta alla realizzazione Atti dell'Accademia dei Georgofili, Anno 2002, settima serie. Industrie Grafiche Fiorentine, Firenze.
77. Pacciani A (2002) Funzioni sociali dell'agricoltura e nuovi strumenti di intervento pubblico XXXIX Convegno SIDEA. "Nuove tipologie dell'impresa nell'agricoltura italiana", 12–13 settembre, Firenze.
78. Pacciani A (2003) La Maremma Distretto Rurale: un nuovo modello di sviluppo nella consapevolezza della propria identità. Editrice Il mio Amico. Grosseto
79. Pacciani A Toccaceli D (2010) Territorio, imprese e istituzioni nella PAC oltre il 2013: l'agricoltura grossetana di fronte a nuovi scenari, In:Pacciani A. Toccaceli D. (Eds), Le frontiere dello sviluppo rurale. L'economia grossetana tra filiere e territorio Milano: Franco Angeli ISBN: 978-88-568-3268-6
80. Pacciani A., Toccaceli D (Eds) (2014) Percorsi di governance per la valorizzazione delle aree rurali nella prospettiva di riforma delle politiche europee I Georgofili Quaderni 2013-II Società Editrice Fiornetina, Firenze ISBN: 978-88-6032-294-4
81. Pecqueur B (2001) Qualité et développement territorial: l'hypothèse du panier de biens et de services territorialisès. Économie Rurale. 261: 37-49. doi: 10.346/ecoru.2001.5217,http://www.persee.fr/web/revues/home/prescript/article/ecoru_0013-0559_2001_num_261_1_5217
82. Perrier-cornet P., Sylvander B (2000) Firmes, coordinations et territorialité. Une lecture économique de la diversité des filières d'appelation d'origine In: Economie Rurale 258:78-89, pp78-89. DOI: 10.346/ecoru.2000.5195. http://www.persee.fr/web/revues/home/prescript/article/ecoru_0013-0559_2000_num_258_1_5195
83. Petrovich B, Manzoni P (2010) I distretti nella regione Lombardia. Rete Rurale Nazionale, Roma.
84. Porter M. E. (1998) Clusters and the new economics of competition In: Harvard business review 76(6):77–90. ISSN 0017–8012
85. Porter E (2003) The economic performance of Regions, In: Regional Studies, Vol. 37.6&7, pp549-578, August-October DOI: 10.108/0034340032000108688

86. Porter M., Ketels C (2009) Clusters and industrial districts: common roots, different perspectives in Becattini G., Bellandi M., De Propris L. (Eds) A handbook of industrial districts, Edward Elgar, Cheltenam UK, Northampton, MA, USA:172–183 ISBN : 978-1-84720-267-3
87. Porter M.E., Ketels C., Miller K., Bryden R (2004) Competitivness in rural U.S. Regions: Learnings and research agenda, U.S. Economic Development Administration, February 25, 2004 http://www.hbs.edu/faculty/Pages/item.aspx?num=46871
88. Pyke F., Becattini G., Sengenberger W. (eds) (1990) Industrial districts and inter-firm Co-operation in Italy, IILS, ILO, Geneva
89. Ray C (1998) Culture, Intellectual Property and Territorial Rural Development. In: Sociologia Ruralis. , . doi: 10.1111/1467-9523.00060
90. (1999) Il distretto agroindustriale della Marsica. Grafiche Di Censo, Avezzano.
91. Romano D (2000) I sistemi locali di sviluppo rurale CNEL, Secondo rapporto sull'agricoltura. L'agricoltura tra locale e globale. Distretti e filiere. Collana Documenti CNEL, II serie. N.23, Roma
92. Rouger M.P. (2014) Pays de fougeres ECOBATYS: De l'observatoire à la demarche de prospective In: Pacciani A., Toccaceli D (Eds) Percorsi di governance per la valorizzazione delle aree rurali nella prospettiva di riforma delle politiche europee, pp51-60, I Georgofili Quaderni 2013-II Società Editrice Fiornetina, Firenze ISBN: 978-88-6032-294-4
93. Saraceno E. (1992) Il distretto delle barbatelle. La Questione Agraria, n. 46, pp.209-231 La questione agraria 46:209-231. Franco Angeli, Milano, ISSN: 1593-8441
94. Sforzi F (1987) L'identificazione spaziale In: Becattini G. (ed.) Mercato e forze locali: il distretto industgriale, Il Mulino, Bologna.
95. Sforzi F., Mancini M. C. (2012) A reinterpretation of the agri-food system and its spatial dynamics through the industrial district In: Arfini F., Mancini M.C., Donati M., Local agri-food systems in a global world: market, social and environmental challenges, pp 9–27, Cambridge Scholars Publishing, Newcastle upon Tyne, UK ISBN (10): 1-4438-3664-8, ISBN (13): 978-1-4438-3664-7

96. Spampinato D., Timpanaro G (2012) La competitività dei sistemi produttivi territoriali in Sicilia: alcuni casi studio per un'analisi dei distretti agroalimentari in Economia e Diritto Agroalimentare n.1 Economia & diritto agroalimentare XVII: 55-82. University Press, Firenze
97. Tarangioli S (2008), Rural and quality food districts as governance instruments: the Italian experience. WP presentato alla Conferenza Europea "Europe's rural areas in action. Facing the challanges of tomorrow. Cipro, 16–17 ottobre
98. Tarangioli S (2011) La progettazione integrata nelle colline del Medio Friuli, in: Seminario Nazionale su Politiche, istituzioni e sviluppo rurale:come migliorare i processi di governance. Rete Rurale Nazionale, Roma.
99. Tinacci Mossello M (2002) Rischi di Crisi e Nuove Opportunità di Sviluppo per i Mercati Locali nel Quadro Della Globalizzazione. In: Rosa M, Vincenzo D (eds) Tra globalizzazione e localismo, Liguori Editore, Napoli. pp 33-50
100. Toccaceli D (2006) Il Distretto rurale della Maremma: 1996–2006. Come si forma un Distretto rurale in Agriregionieuropa, Anno 2, n. 6 pp. 54-56, ISSN: 1828 - 5880
101. Toccaceli D (2012) Dai distretti alle reti? I distretti in agricoltura nell'interpretazione delle Regioni e le prospettive verso il 2020. Rete Rurale Nazionale, Roma.
102. Toccaceli D (2014) Sustainable Development in Rural Areas: A Comparison on Governance Among Some European Experiences. In: Gennaro BC, Nardone G (eds) Sustainability of the Agri-Food System: Strategies and Performances, Proceedings of the 50th SIDEA Conference, Lecce, Chiostro dei Domenicani, Universitas Studiorum, Mantova. pp 307-321 ISBN 978-88-97683-60-5
103. Tregear A., Arfini F., Belletti G., Marescotti A (2007) Regional foods and rural development: The role of product qualification In: Journal of Rural Studies, Vol. 23, Issue 1, Pages 12-22, Elsevier Ltd, doi:10.1016/j.jrurstud.2006.09.10

Chapter 4

Duty of Stewardship and Fisheries Governance: A Proposed Framework

Adam Soliman

The Fisheries Law Center, 183 – 8623 Granville Street, Vancouver V6P 5A2, BC, Canada

ABSTRACT

Fisheries often fall prey to overfishing and the exhaustion of stock. Fishing governance is an ongoing attempt to prevent such an outcome. Over time, fisheries regulation has generally moved from controls on inputs to controls on output, such as catch limits and Individual Transferable Quotas. Individual Transferable Quotas have reduced overcapitalization, and have in some cases allowed stocks to rebuild. However, because they enable market trading of catch shares which tends to concentrate fisheries in fewer hands.

This paper proposes applying a duty of stewardship to the existing fisheries governance structure. "Stewardship" is an obligation to be responsible for taking care of another person's property. The concept of stewardship easily applies to fisheries, because fisheries are natural resources which belong to the public. Current regimes, such as Individual Transferable Quotas ("ITQs") do not do enough to prevent the employment of destructive fishing practices and place the burden of natural resource management on the government. Assigning a duty of stewardship upon fishers, whether they own or lease an ITQ, would require fishers to be stewards of common resources and use responsible fishing practices.

INTRODUCTION

The *tragedy of the commons* is a well-known phenomenon that describes what can happen if there are no controls on the use of a shared resource and too many users willing to deplete that resource (Hardin [1968]). Fisheries often fall prey to the tragedy of the commons, and fishing regulation can be regarded as the ongoing attempt to prevent this outcome (Macinko and Schumann [2008]). A variety of methods have been employed to manage fisheries, from input controls such as designated season lengths and types of gear that can be used to output controls such as catch limits. (Macinko and Schumann [2008]). Many countries have implemented Individual Transferable Quotas, a form of property right that can be transferred. While ITQs have reduced overcapitalization of fish stocks, they have also enabled market trading which concentrates fisheries in fewer hands (Ecotrust Canada [2009]; Bromley and Macinko [2007]).

This paper proposes applying the duty of stewardship to all fishers. ITQs have reduced overfishing, but have insufficiently protected fisheries from destructive fishing practices and have resulted in the concentration of fishing into the hands of a few with less reason to think beyond the profits garnered today. As it stands today, the regulator alone is accountable for protecting the resource. Therefore, it has become necessary to legally require that fishers be responsible for protecting the health and productivity of the resource from which they extract. If the immediate benefits of employing certain types of fishing techniques outweigh the negatives of using other practices which are better for the long-term, then fishers have no incentive to be stewards of the fisheries. Consequently, the duty of stewardship must be applied to fishers via individual fishing contracts and/or government regulation that can be applied to any fisher.

FISHERIES REGULATION: CONCEPTS AND HISTORY

The tragedy of the commons theory suggests that multiple individuals, acting independently and rationally on behalf of their own self-interest, will ultimately deplete a shared limited resource if there are no individual property rights. According to the theory, this will happen even though each individual knows that depletion of the resource is not in their own interest, nor in anyone's long-term interest (Hardin [1968]).

Private property rights are one way to effectuate sustainable stewardship of shared resources. Farmers are incentivized to conserve land when given permanent and exclusive ownership of that area (Leal [2000]). When the owner can reap the benefits as well as suffer the consequences for misuse of property the owner is incentivized to consider the long-term consequences of mis-management land use. As Ziff puts it: "Private property can operate to reduce the ability of an owner to shunt off costs onto others. In the language of economics, private ownership can reduce social costs or 'negative externalities'" (Ziff [1996]). While private property incentivizes more sustainable stewardship of common resources, private ownership does not guarantee conservation of resources. If the value of short-term exploitation outweights the value of long-term conservation, then economic theory suggests that private owners have no incentive to act in such a way to protects the future availability of that resource. In fact, most private owners prefer profits now over profits later (Macinko and Bromley [2004]). Bromley, presents several examples of laws that have been promulgated in order to legally require that property owners take care of their property, and argues that these laws would not need to exist if ownership was a sufficient incentive to maintain that property in good order. He concludes: "To be precise about the matter, if the 'time preference' of a private owner is such that income now trumps income in the future, then private owners will be quite intent on liquidating (destroying) a renewable natural resource in order to spend the proceeds—or invest them elsewhere" (Bromley [2009]). Consequently, private property rights are not always sufficient to prevent the destruction of a shared resource.

Private ownership is only one solution to natural resource protection. Regulations which focus on the harm can also serve to preserve shared resources. As Ziff points out, "the tragedy of the commons story is not so much an argument in favour of private property as it is a manifestation of the perils of *unregulated* common property" (Ziff [1996]). Many researchers do not believe that the solution lies in property laws. Driesen, for example, advocates for a solution that focuses "upon the harms" and not on ownership (Driesen [2002]). For example, pastures in the Swiss Alps have successfully been managed as common property for centuries without suffering from overgrazing. (Ziff [1996]; Taylor [1992]; Stevenson [1991]). Another solution includes a permit system which places limitations specifically on the permit holder. In summary: private property rights are but one one possible way of preventing the tragedy of the commons. Researchers hold widely varying views about the relative effectiveness of the various possible solutions. What is not in doubt is that when there is completely unregulated, open access to a limited shared resource, then that resource is very likely to be depleted and destroyed (de Schutter [2011]). Without regulation, natural resources like fisheries are likely to be depleted.

Tragedy of the Commons in Fisheries

The characteristics of shared fish stocks make them vulnerable to the tragedy of the commons. In the absence of regulation, each fisher owns only the fish he can catch (Churchill and Lowe [1999]). Under English common law, this is known as the *rule of capture* (Scott [2000]). Given that most fish stocks do not stay in one place, a fish which is not caught today may not be available to fish tomorrow. This promotes the immediate depletion of the fisheries in order to take advantage of their benefits today (Churchill and Lowe [1999]). Eggert and Ulmestrand ([2007]) describe the situation as follows:

The problems related to open access marine fisheries lead fishermen to lack in responsibilities for future benefit flows. In fact, theoretically, open-access fishery implies that fishermen only care about their own catch today, but completely disregard their potential catch tomorrow and in the more distant future (Clark [1973]). It is sometimes held that wealth or the prospect of wealth leads to over exploitation of resources and that myopic behaviour of fishermen is a result of the human nature (Ludwig et al. [1993]). However, from an economics perspective, the problem is rather that poorly defined property rights lead to perverse economic incentives, which causes excessive effort and shortsightedness (Björndal and Munro, [1999]).

Ownership alone is insufficient to prevent the irresponsible depletion of shared resources. Ostrom finds that the tragedy of the commons applies to fisheries with only one significant difference: the basis for ownership is capture rather than long-term possession. Coastal waters are ripe for overfishing, but regulatory controls and enforcement of such controls can protect fisheries.

Fisheries Regulation through Input Controls

Among the first effort to preserve fisheries was the introduction of licences (Squires [1995]). Rather than having open access to fisheries, fishers now had to have licenses which granted them the right to fish. Only a limited number of licenses were available to be given out. As fishing technology has improved, fishers have been able to vastly increase their catch. This meant that the total fishing capacity of the fleet came to easily surpass the sustainable yield. In response to overfishing, governments applied strict conditions on licences, but problems persisted (National Research Council [1999]). In some instances, fishing was limited to a certain number of days per year. In other instances, licenses limited the size of the vessels that could be used. Unfortunately these limitations had little effect – fishers quickly developed more efficient ways to fish the same amount of fish in the reduced timeframe and enhanced their fishing nets to make up for the size of their boat. Without protections, fishers had no incentive to fish sustainably because if they did not catch the fish today, their competitor would catch that fish.

To combat overfishing, governments restricted fishing by setting limits on the quantity of fish stocks that could be fished by license holders (Kerr et al. [2003]). Thus, the regulatory approach shifted from a system by which who could fish and what could be used to fish was controlled (i.e. input control) to a system where catch limits were imposed (i.e. output control). Many output control regimes use a type of limited property right called an Individual Transferable Quota ("ITQ"). In an ITQ regime, the government sets a Total Allowable Catch (TAC) for a region and/or a fish species, and grants fishers shares of the TAC, what is known as an ITQ. ITQs are transferable – they can be leased and sold. Transferability is permitted becauseit is expected that this will lead to "efficiency"; the fishermen with the lowest costs will be able to place the highest bids for the ITQs (FAO [2004]). Herein lies the problem – the license holders no longer have any more than a financial stake in the health and viability of the fishery.

ITQs have reduced overfishing and helped to preserve fish stocks. The most important impact of ITQs has been to put a stop to the "race to fish." Some fisheries utilize the TACs, but do not assign individual quotas. This means that each season a TAC is set and once that TAC has been met in the aggregate the fishing season is closed. This system incentivizes individual fisheries to use whatever techniques to capture as many fish as quickly as possible. There are no consequences for using techniques that are harmful to the health of the fishery (such as techniques that capture significant amounts of bycatch in addition to the desired catch). Thus, behaviors dedicated to simply maximizing profits do not guarantee the most rational behavior nor the most productive, sustainable approach. Market based principles such as profit maximization do not actually guarantee that industries will act in their best interests.

While ITQ regulations have succeeded in conserving the particular fish stock managed by the ITQ regime, there is no certainty that ITQs alone can protect the future viability of the fishery (Chu[2008]). If the TAC for a fishery is set too high, ITQs may succeed in limiting the catch to that amount, but too many fish will be caught and the resource will eventually be exhausted. Therefore, ITQs make little difference unless the TAC is set at the appropriate level. But even if the TAC has been properly set, exhaustion can be economically rational under a property rights regime if the benefits of fishing today outweigh the benefits of leaving fish for tomorrow (i.e. the discount rate). Sumaila modeled Canada's Atlantic cod fishery off Newfoundland, and found that a discount rate of 20% resulted in a predicted collapse that closely paralleled the actual collapse that occurred after ITQs were introduced to that fishery (Sumaila [2010]).

ITQs have been criticized for a number of reasons. First, ITQ regimes have enabled fishers with licenses to reap significant profits, but have not set aside any money for the public or for the protection of the natural resource (Bromley [2009]). As a public resource, it seems reasonable that the public should be compensated for the yearly depletion of the fish, even if it is sustainably depleted. Also, it would seem that license-holders should be responsible for contributing to efforts to maintain the health and viability of fish and their habitat. Second, ITQregimes typically concentrate ownership and profits into a few parties, usually large fishing companies and processors (Copes and Charles [2004]). This often results in job losses and can be associated with the disappearance of small fishers who cannot compete with the larger companies. The hardest hit by the concentration of licenses in large fishing operations are the coastal communities and small-scale fisheries who depend on the fish for a livelihood and food (Davis [1996]). As a result, ITQs have often faced intense opposition from small-scale and traditional fishers (Macinko [2007]). Researchers are concerned about the consequences of utilizing ITQs in Third World nations, where traditional fisheries are seen as necessary for the survival of millions of low-income people (Kurien[2004]). Even in wealthier nations, there are fears about the long-term impacts on ecosystems and communities that may result if economic efficiency is allowed to be the sole determinant of fisheries policy. Consequently, many are seeking alternative policy strategies that could achieve some of the benefits of ITQs while avoiding their worst negative impacts.

AN ALTERNATIVE: THE DUTY OF STEWARDSHIP IN FISHERIES

This paper proposes defining the legal duty of stewardship to fisheries and marine ecosystems and applying this duty of stewardship to the existing fisheries governance structure.

What is a *duty of stewardship*? Black's Law Dictionary defines a steward as "a person appointed to manage the affairs of another" (Garner [2009]). Thus, the person who owns a valuable item outright generally has no duty of stewardship with respect to that item. There is nolegal obligation to take care of the item owned, though there may be obligations to take care that your ownership does not affect other privately owned items. Duties of stewardship, either implicit or explicit, abound in our everyday lives.

Land is one privately owned "item" where many researchers have discussed the application of the duty of stewardship. Legal experts, such as Hamilton ([1998]) and Karp ([1993]), have argued that landowners should be subject to a duty of stewardship with respect to their land, even if they own it in *fee simple*. Under traditional concepts of land ownership, a landowner has complete freedom to use their land as they wish. This view is contrary to traditional concepts of ownership, which are founded on the idea that a landowner has nearly complete freedom to do as he likes with his land. This traditional view, however, is severely weakened by the fact that what an owner does with his land can have significant impacts on other people, across space and time. According to Karp ([1993]), "[l]and is fundamentally different from other forms of property. Because any parcel of land is part of a network of natural systems extending beyond the boundaries described in the deed, it attains an importance superior to any individual landowner or to any period in time. Land is essential to our right of survival". In a few jurisdictions, notably Iowa, this concept has been recognized and validated in common law, thus restricting the freedom of landowners to some degree (Hamilton [2011]).

The concepts of stewardship and "land ethic" are highly controversial in the agricultural realm because these concepts appear to infringe on the presumption that the owner of real property should be free to do with their property as they wish. Legally speaking, fisheries are different from farms. While land owners own the land, fishers own neither the uncaught fish, nor the water in which the fish swim. Uncaught fish and marine resources, generally, are considered the property of the public. Therefore, the concept of a duty of stewardship – a responsibility to take care of the property of others – is compatible with the legal nature of a fishery. Because fishers are not owners, fishers must be explicitly subject to a duty of stewardship in order to be held legally accountable for fishing responsible.

Most proponents of ownership-based schemes such as ITQs argue that ownership itself promotes the conservation and protection of marine resources similarly to someone subject to a duty of stewardship. ITQ supporters believe that proper stewardship behavior results from a properly designed and implemented ITQ regime (Bromley [2009]). Macinko and Whitmore dismiss this claim, and state: "The simple truth is that ownership does not ensure stewardship or conservation. We know ... that both private and public owners may either care for natural resources or abuse them" (Macinko and Whitmore [2009]).

While ownership may or may not improve the odds that an individual will be a good steward of natural resources, another important question remains. Will the future stream of benefits derived from being a good steward exceed the current benefits of natural resource overexploitation? Or in other words, what is a fisher's incentive to be a steward? It is rational to expect an owner to exploit natural resources if the benefits of taking natural resources today is greater than the benefits of ensuring their future existence. Much of this depends on the actions of other fishers who are at the same time weighing the same concerns. The fear is that another fisher will get what you did not catch, but have the capacity to catch. Therefore, without legislated or contractually set limitations on how and what a fisher can catch, the tragedy of the commons persists.

Poaching and illegal, unreported, unregulated fishing (IUU) alters the benefits that fishers can expect in the future. If fishers believe that other fishers can cheat and increase their returns, the future benefit stream will be seen as risky. Effective and credible enforcement will reduce the risk of being a good steward and balance current and future benefits. Therefore, proper enforcement is an integral part to the solution. Inherently this means that the management scheme would not in fact be voluntary. For fisheries, stewardship does not automatically flow from ownership. Consequently a duty of stewardship needs to be explicitly created and imposed (Macinko [2007]). It is dangerous to assume that adequate stewardship will result simply from assigning ownership rights.

According to Macinko stewardship should be the goal itself, a mechanism for achieving conservation. He describes stewardship as "an ethic, an attitude, a responsibility" (Macinko[2007]). For Macinko, the term "stewardship" represents the legal and operational mechanism that this paper calls a "duty of stewardship", and the underlying meta-governance principles and ethics that justify creating this mechanism.

Ownership of property is tied to certain property laws, depending on the type of ownership. Without a legislated or contractual duty, the duty of stewardship is to be understood as a duty under tort law. The idea is that the duty of stewardship is applicable regardless of the category of property rights at play or it can be utilized in association with management measures that do not involve property rights. The duty of stewardship is intended to supplemental existing fisheries management methods because the current schemes are incomplete. It is not a standalone duty.

THE UNDERLYING META-PRINCIPLES

The application of a duty of stewardship is further justified by meta-principles or the underlying principles and assumptions of fisheries governance (Kooiman et al. [2008]). As Kooiman et al. ([2005]) says, "[w]hen governors define the problems they think should be addressed and the solutions to these problems, they inevitably draw on fundamental assumptions and worldviews that should be brought to the surface so they can be explained, defended and examined". Kooiman proposed a list of twelve ethics and principles for the meta-governance of fisheries (Kooiman et al.[2005]). These principles justify the application of a duty of stewardship to fisheries governance. According to the meta-principles, the preservation of species and ecosystems is morally good. A duty of stewardship promotes environmental ethics by requiring fishers to preserve the marine environment on which they depend.

The duty of stewardship has been linked to the "land ethic" and sheds light on the application of the duty upon actors within fisheries. Hamilton ([2010]) wrote:

[I]n many ways Leopold's views provided a theoretical underpinning for the modern ecology movement, which attempts to alter how man relates to the environment and to assign the costs of environmental degradation to those responsible... Even if the United States experiences difficulty developing a true Leopoldian land ethic, it does not mean society is without methods to address agriculture's impact on the environment. In an increasingly legalized society we have come to rely on laws and legal duties as a substitute for a land ethic. Laws which impose a duty on landowners to protect their soil from erosion, such as one enacted in Iowa in 1971, and new laws requiring farmers to account for their impact on the environment have the effect of changing the man-land relation. It is true, by regulating the relation of individuals to society rather than the individual to the land, these laws function in a different ethical dimension than would a true "land ethic." But as society comes to view protecting the environment as a significant societal goal, the substitution of legal duties, although not a perfect proxy for a "land ethic," may be the legacy of the second stage of environmental policy toward agriculture.

The purpose of the "land ethic" is to obligate individuals to be responsible stewards of the land and prevent land degradation. Hamilton does not analyze the success of Iowa's laws in preventing land degradation. However, Eswaran argues that few jurisdictions in the world are successful in preventing land degradation. In addition, there is considerable disagreement about definition of land degradation, and about how it should be assessed and measured (Eswaran et al. [2001]).

The scope of a duty of stewardship driven by the considerations of environmental ethics will vary depending on the focus of the ethic. Under a human-centred environmental ethic, organisms other than humans are valued only according to their effect on humans. A humancentric focus in turn narrows the scope of the duty of stewardship wherein the obligation to prevent environmental degradation extends as far as harms that negatively impact humans. Under a nature-centered environmental ethic, all organisms are morally relevant. Consequentially, the scope of the stewardship duty would be broader because actors would be obligated to avoid negatively impacting any nature.

Meta-Principle: Sustainability

Examining the application of the meta-principle of sustainability to agricultural practices also helps to shed light on the relevance of applying the duty of stewardship to fisheries. Practitioners debate the definition of "sustainability," but at its simplest it can be described as meeting the needs of the present without compromising the ability of future generations to meet their own needs (National Academy of Sciences [2010]). In the context of agriculture, Hamilton ([1998]) provides the following highly informative definition:

Sustainable agriculture is defined in various ways, but in its simplest form, it means developing agricultural practices which protect the environment while preserving the economic profitability of farmers. The basis of the concept is that no agricultural system can be successful in either the short or long term unless it is designed to sustain the resources necessary for its operation. These resources include both physical resources of soil, air, and water, and also human and social resources of farm families, rural communities, and the economic structurenecessary for an agrarian system to function.

In other words, sustainable practices provide for current use of natural resources which is not at odds with the future health and viability of those natural resources. Iowa law defines "sustainable agriculture" as actions "preserving the high productivity and quality of Iowa's land" (Iowa Code[1997]). Under Iowa law, agriculturists are required to care for the land so that it remains productive for someone else's use in the future (Iowa Code [1997]). This legal requirement can be equated to the type of duty of stewardship advocated to be applied to fisheries – it is an obligation to preserve a natural resource, may it be land or a fishery, in good enough condition that a future user can enjoy the same quality of resource.

Fisheries are similar to, but not the same as land and agriculture. In the case of fisheries, the general public owns the resource, not the fishers. Even when fishers own quotas, that quota represents a beneficial right to catch a certain quantity of fish not a specific quantity of uncaught fish in the ocean that they own. Sustainable fishing is fishing in a manner that maintains and protects the ecosystems and populations needed to sustain the fishery. Because fishers do not own the fish or the ecosystem in which the fish live, the obligation to be stewards and the benefits derived from being good stewards are not necessarily self-apparent. However, in order to maintain livelihoods and food sources that have been depended on for decades, perhaps even centuries, the responsibility to preserve the fisheries becomes more obvious.

Meta-Principle: Food Security

Another important meta-principle and justification for expecting/ requiring fishers to be stewards is food security (Bavinck and

Chuenpagdee [2005]). Food security is sometimes regarded simply as a matter of quantity, but this is insufficient. The World Bank defines food security as "access by all people at all times to enough food for an active healthy life" World Bank ([1996]). Thus, achieving food security requires ensuring that an adequate food supply is physically and economically available to everyone, especially in regions that cannot produce enough food because of climate or geographic constraints, and especially for people with low incomes (Chuenpagdee et al. [2005]). Not surprisingly, small-scale fisheries have been recognized as playing a vital role in improving food security in many countries (Kooiman and Jentoft [2005]).

A duty of stewardship will protect small-scale fisheries and their ability to serve as a livelihood for many and food for others. Small-scale fisheries are the fishers most at risk of unsustainable fishing and illegal fishing. Obligating all fishers to be stewards of the fishery and surrounding ecosystem will benefit everyone, the fisheries and the communities who depend on fishing for a livelihood and source of food.

The Precautionary Principle

The duty of stewardship is essentially a call to not act without knowing the consequences. The precautionary principle states that one should not undertake an activity for which there is no evidence of its safety. Specifically, within the environmental context, the principle states that when ecological or health risks are suspects, those risks should be reduced and less-risky alternatives should be pursued where possible (Myers and Raffensperger [2006]). Even in the face of inconclusive science, precautions should be taken. Furthermore, before engaging in activities that could potentially harm the environment, the risks should be researched. As good stewards, the mere possibility that an activity could cause irreparable harm would be sufficient to prevent further actions.

Kooiman recommended that the precautionary principle be applied to fisheries governance in the Code of Conduct for Responsible Fisheries (FAO [1995]). According to the FAO: "[m]anagement according to the precautionary approach exercises prudent foresight to avoid unacceptable or undesirable situations, taking into account that changes in fisheries systems are only slowly reversible, difficult to control, not well understood, and subject to change in the environment and human values" (Kooiman et al. [2005]). It is hard to say how the actions of today will affect the viability of the fisheries in the future. Consequently, it is in the best interests of fishers to investigate any potential risks from activities before engaging in them. The benefit of requiring the application of the duty of stewardship via legislation or individual fishing contracts is that if a steward has any question about the safety of his/her fishing practices and harms result as a result of his/her fishing, he/she would be responsible.

Meta-Principles: Responsiveness, Respect and Inclusiveness

Kooiman also proposes applying the principles of responsiveness, respect and inclusiveness to fisheries governance (Kooiman et al. [2005]; Kooiman and Jentoft [2009]). All fishery stakeholders should be provided the opportunity to be heard and have their interests protected. They should be actively engaged in the creation and implementation of a fisheries governance structure, which should be structured in such a way to respond to concerns. By including all the stakeholders in the governance structure, you are more likely to get engaged stakeholders who understand and comply with the rules. Furthermore, the group can take advantage of the myriad of expertise available to them via the participation of such a wide group of stakeholders. Many stakeholders already regard themselves as stewards and their leadership within these governance structures can help to educate others.

The Public Trust Doctrine

The Public Trust Doctrine states that the government holds certain natural resources, such as navigable waters, in trust for use by the public. Brewer and Libecap ([2007]) explain the public trust doctrine in greater detail:

To insure group values are respected, the rights of the public are vested in the state as trustee of the resource. As such, the state through its administrative agencies has a duty to administer, protect, manage and conserve the resource. Any existing private users have only usufruct rights that can be withdrawn whenever the state deems that they are inconsistent with the public trust. Because there are no private property rights, there is no basis for takings challenges in such reallocations. Furthermore, the legislature cannot alienate trust resources, which must remain with the state.

Under this doctrine, the State is the steward for resources on behalf of the public. In addition, the doctrine holds that the public's ownership of the resource cannot be alienated, and so the resource cannot become private property. If the government were to deny access by the public to these resources held in trust, a "takings challenge" could be initiated against the government (Hamilton[2010]).

A large shortcoming of this doctrine is that while the government has self-designated as the "steward" of such resources, the only actions they are required to take are those that are explicitly prescribed by law. Otherwise, the government is not obligated to act to steward the resources. Therefore, a government cannot be sued unless a statute requires the government to do something, and the government fails to do it. In short, even though the doctrine places the state in a stewardship role, it does *not* impose a general requirement that the government behave responsibly.

Usufructuary Obligations

Usufructuary obligations are similar to that of a steward's as described here. A usufruct has the right to enjoy the property and the fruit of that property (i.e. the profits that can be derived from the property such as crops), but does not own full rights to the property. However, the usufruct can lose their right to the property if their actions destroy the property. According to LeVan ([1962]), under Anglo-American common law, a legal life tenant "may not change the premises in such a way that remaindermen or reversioners would have "reasonable ground for objection thereto". In Japan, fisheries are managed by Fishery Cooperative Associations ("FCAs"), organizations made up of local fishers (Lim 199–200). The FCAs are responsible for designing the rules their members follow and coordinating with the government which sets allocation limits (McIlwain [2013]). A usufruct prescribes essentially the same obligations as the duty of stewardship. A usufruct is allowed to utilize the resources available to them on the property;

however, they cannot reduce the value or quality of the property or natural resources on the property.

INTRODUCING STEWARDSHIP IN FISHERIES

Assigning ITQs is just part of properly managing fisheries. ITQs must be supplemented with the application of a duty of stewardship in order to ensure that fishers are required to sustainably fish. Fishers can be motivated to become stewards through legal requirements, contractual requirements or consumer expectations. And for any of these structures to work it is imperative those being managed, the regulators as well as the consumers must understand what is required to be a "steward."

Small-scale fisheries in developed countries are recognizing the importance of differentiating their products through sustainability. But to achieve this distinction requires an underlying and fundamental attitude change: these fishers must be able to present themselves as stewards of the sea. In order to benefit from this distinction fishers have developed various marketing schemes such as Consumer Supported Fisheries (CSFs) and Sustainability Labeling Schemes that highlight the use of sustainable fishing methods and the compliance with any fishing regulations in place. These schemes use tools such as eco, sustainable, and wild labeling, and traceability techniques to connect consumers with fishers (Kooiman and Jentoft [2005]). Such marketing will entice consumers to buy only sustainably caught fish and fish-products, thereby pressuring other fishers to become stewards.

The effective use of market-based incentives such as special labels or preferred purchasing agreements requires both a voluntary adoption of the duty of stewardship and an educated consumer. First, consumers must understand how important it is that fishers are stewards. They must be educated as to the import of fishing sustainably and complying withlimits on the stocks and areas in which they can fish. If consumers understand this, they will be more inclined to purchase products to be fished sustainably. Second, fishers must benefit from the voluntary adoption of stewardship practices. In other words, fishers must reap benefits from being distinguished from fishers who do not fish sustainably or act as stewards. While market-based incentives may be effective, the circumstances call for a more immediate shift in practices. Legislation must be promulgated that requires fishers to be stewards, or contracts must be revised/written anew to include provisions requiring stewardship by license holders. There are many fishers who currently, without regulation, act as stewards.

Unfortunately the vast majority of fishers may fish within the limits of their ITQs, but not act as stewards. Being subject to a duty of stewardship would mean that a fisher who uses or consumes a fish stock in a way that destroys that stock or its habitat could be held liable for the damage he/she caused. Explicitly associating a duty to conserve the stock with holding a quota or licence would allow the possibility of legal action by a third party against a resource user who neglects this duty – not only against the regulator, as in the example in the preceding section. If enforced, fishers would be motivated to act as responsible resources managers.

ITQs have presented one of the largest challenges to promoting stewardship because ITQs can be leased, which has been show to weaken the incentives:

According to some fishery managers and experts, leasing reduces stewardship incentives, which may impact the community's long-term economic viability. Quota leasing separates the person holding the quota from the person fishing the quota. In some cases, quota leasing may diminish stewardship incentives by creating a class of absentee quota holders who rely on independent fishermen. While owner-on-board rules, such as those in Alaska, may minimize the risk of creating this class of absentee quota holders, fishermen who lease quota have only a temporary privilege to catch fish. Thus, they have less interest in the long-term health of the fishery, especially as the end of their lease term approaches. Consequently, incentives may exist to catch more fish than their quota allows and sell this over-quota fish on the black market or to fish using nonsustainable methods. For example, according to New Zealand fishery experts, quota holders in the high-value abalone fishery found that unskilled fishermen who leased quota were jeopardizing the fish by extracting them in ways that harmed the abalone beds. (Government Accountability Office [2004])

When ITQs are leased, the lessee is less likely to have any feeling of obligation to do more than take advantage of their current license to fish. But the lessor has an incentive to ensure that the lessee does in fact act as a good steward. Similar concerns about the lack of an incentive for leaseholders to conserve a resource arise in agriculture. To combat that, many farm lease agreements include a "good husbandry provision", which act to impose a stewardship obligation upon the leaseholder. Retired landowners in Iowa are increasingly choosing to impose such a duty in farmland leases ([Cox 2011]a). The Iowa Supreme Court has held that "the purpose of a good husbandry provision is mandating the proper use of land rather than requiring high yields", a finding that clearly creates an obligation to act as a good steward of the owner's land ([Cox 2011]b).

If the duty of stewardship was acknowledged as an obligation on all fishers, including the lessees of ITQs, the lessor would be able to hold the lessee accountable for any damages caused (Hamilton [2010]). Where fishers use destructive practices in instances where they could use lessdestructive fishing gear or methods, lessor could get damages. It is in the lessor's interest to safeguard the future stream of benefits, while the lessee is exposed to strong incentives to maximize his current benefits (Munro et al. [2009]). The duty of stewardship allows the lessor to do so even in the absence of strong enforcement. It might be necessary to state some of the practices that are unacceptable to the lessor, but it would be impossible to include an exhaustive list of everything that is and is not acceptable. Therefore, it will probably be necessary to definethe duty of stewardship in such a way that allows for unforeseen situations in addition to examples of known unacceptable practices. The benefit is that although the duty would cover a wide range of situations, the inclusion of examples would guide the courts and reduce the risk that a court might find the clause to be too vague to be enforced.

A stewardship clause in a quota lease agreement could resemble the following: "It is the duty of the lessee to exercise good stewardship with respect to the fish stock and the environment in which the fish stock lives. Good stewardship is defined as neither fishing, nor conducting other activities in the marine environment, in a manner which will or is likely to significantly degrade the fish stock or its environment, according to experience and/or scientific knowledge that is reasonably available to fishermen, UNLESS no reasonably feasible alternative manner of operation exists." There has been significant research on the proper methods of fishing.

Therefore, this research would serve as the basis for determining what practices and actions do and do not qualify as good proper stewardship. For example, if a lessee was found to utilize bottom trawling which has been found to destroy coral reefs and take in bycatch, the lessee would be found liable for failing to fulfill his duty of stewardship.

At the same time that regulations are put in place, economic incentives can also be used. For example, in order to be able to renew a fishing license the license holder would have to demonstrate adequate stewardship. If a fisher has *not* been found in breach of this duty, his/her license or quota could be renewed; otherwise it could be revoked. This would also mean that if the ITQ holder leases the quota to another fisher and that fisher fails to comply, the license holder could lose the license. This in essence disincentivizes the licensing of ITQs (which has other negative impacts on fishing, including pushing out small-scale fisheries) and makes the licensor accountable for properly vetting and monitoring the actions of the licensee. One of the benefits is that regulators would not have to dismantle the entire scheme to punish one person. Instead, the regulator would have the authority to cancel a specific fisher's quotas without affecting the system as a whole.

Ideally countries will implement a legislative duty of stewardship. However, if neither statutory law or a contract explicitly defines a duty of stewardship, the courts may be asked to find whether an implied duty of stewardship exists under common law. Different jurisdictions may answer this question differently, unless a higher court with a broader jurisdiction has made a decision. It is then up to practitioners to ask a court whether a duty exists, and whether a party has breached their duty. This is what happened in *Heiderscheit*, where Iowa's appeal court found an implied covenant of good husbandry (Heiderscheit [1986]).

The legislature should impose a statutory duty of stewardship by codifying the duty into fisheries acts, using wording resembling that proposed above for a lease agreement.

WHAT OUTCOMES COULD BE EXPECTED?

One of the limitations of systems such as ITQs is that if one or a few fishers damage the resource, the impact is nevertheless absorbed by all quota holders. This may happen, for example, when a TAC must be lowered because of significant overfishing. In short, the "stick" is used against all. In principle, the recognition of a duty of stewardship would make it possible to penalize the specific culprit. This would encourage mutual self-monitoring among the users.

In jurisdictions such as the United States, a duty of stewardship may also have the benefit of replacing the unpredictability and theoretically unlimited scope of tort lawsuits with litigation based on the more specific and limited grounds that are specified by the legislation or precedents defining the fishers' duty of stewardship. A recent lawsuit against the US government for its alleged failure to control bycatch illustrates the problem; it is highly unlikely that the tour boat operator who launched the lawsuit is suffering more than a small fraction of whatever ecological harm has occurred, and it is certainly true that those whose actions could directly cause the problem would not be sued at all (The Boat Company [2012]). A properly defined duty of stewardship would enable actions to be brought on the more relevant grounds of ecosystem and fishery damage, and would enable them to be brought against those who actually caused the damage.

A duty of stewardship will increase inclusiveness in fisheries governance. In catch share schemes, the government authorizes users (quota holders) to use the resource. Other stakeholders are not included in the management framework. For example, conservation groups and environmental non-governmental organizations (NGOs) are often excluded, and can only affect the current system by way of public awareness and other activities outside the management scheme. The public is excluded as well; the government as trustee acts on behalf of the public. Acknowledging the public as a stakeholder will increase accountability by acknowledging thatmore people than just the license holders are affected by the use of unsustainable fishing practices and empowers them to take action.

A duty of stewardship changes the relationship between quota holders and other stakeholders. For example, conservationists, small-scale fishers or NGOs could hold large-scale operators using destructive practices, such as trawlers, accountable. They could sue them for breaching their duty to use least destructive fishing practices. This in turn would allow stakeholders to take an active role in a framework that includes those who were previously unable to participate directly.

In this way, using a duty of stewardship appropriately could change the relationship between government and other stakeholders towards a more inclusive and balanced model. In contemporary fishery management, there is a strong call for self-governance and co-governance, as well as for community-based approaches. The duty of stewardship has the potential to bring the stakeholders into a new form of relationship where the government could itself be held accountable. In other words, the duty of stewardship may level the playing field. Furthermore, to the extent that the duty is shared, it could align the resource users and managers, including the government, so that they are more likely to cooperate in order to act in an ecologically sound manner.

CONCLUSIONS

As the world's human population grows, traditional "open access" capture fisheries have increasingly often collapsed. The "tragedy of the commons" scenario appears to model these fisheries quite well. Governments and regulators have implemented many types of controls in an effort to preserve fisheries. In recent years, ITQs have become a popular solution, partly because it is believed that they provide an incentive to conserve. However, ITQs have been found to concentrate ownership of ITQs in the hands of large-scale fishers and to do little to incentivize conservation.

This paper proposes that supplementing the ITQ regime with a duty of stewardship would require the necessary conservation and preservation needed in order to adequately protect fisheries for current and future use. The duty of stewardship concept originated in the agricultural context, and is well-suited to resolve unsustainable fishing practices. The duty of stewardship could be applied to fishers and regulators through legislation or contractual provisions. The duty of stewardship is also compatible with many important meta-governance principles proposed to be applied to fisheries. Apply an explicit duty of stewardship to fisheries will result in accountable regulators and fishers, will include all the affected stakeholders in the development and implementation the duty as well as the management of the fishery, and result in better fish and ecosystem conservation.

REFERENCES

1. Bavinck, M, and R Chuenpagdee. 2005. Current principles. In Fish for life: Interactive Governance for Fisheries, ed. Kooiman, 237. Amsterdam University Press, Amsterdam.
2. Brewer, J, and GD Libecap. 2007. Property rights and the public trust doctrine in environmental protection and natural resource conservation. Austr J Agri Res Econ 53: 1–17.
3. Björndal, Munro. 1999. The Economics of Fisheries Management: A Survey. In The International Yearbook of Environmental and Resource Economics: 1998/1999, ed. H Folmer and T Teitenberg, 153.
4. Bromley, DW. 2009. Abdicating responsibility: The deceits of fisheries policy. Fisheries 34 (6): 280–290.
5. Bromley, DW, and S Macinko. 2007. Rethinking fisheries policy. Alaska Department of Fish and Game Report. 1–38.
6. Chu, C. 2008. Thirty years later: the global growth of ITQs and their influence on stock status in marine fisheries. Fish Fisheries 10: 217.
7. Chuenpagdee, R, J Kooiman, M Bavinck, P Degnbol, and JJ Pascual-Fernández. 2005. Challenges and concerns in capture fisheries and aquaculture. In Fish for life: Interactive Governance for Fisheries, ed. Kooiman, 303–324.
8. Churchill, RR, and AV Lowe. 1999. The Law of the Sea (Melland Schill studies in international law). Manchester University Press, Manchester.
9. Clark, CW. 1973. The Economics of Overexploitation 181: 630.
10. Copes, P, and A Charles. 2004. Socioeconomics of individual transferable quotas and community-based fishery management. Agric Res Econo Rev 33: 171–181.

11. Cox, E. 2011. A lease-based approach to sustainable farming, part II: Farm tenancy trends and the outlook for sustainability oriented land. Drake J Agric Law 16: 5–29.
12. [http://sustainablefarmlease.org/iowa/] Cox, Edward. 2011b. Sustainable farm lease: a product of the sustainable agriculture land tenure initiative. . Accessed 14 November 2013.
13. Davis, A. 1996. Barbed wire and bandwagons: a comment on ITQ fisheries management. Rev Fish Biol Fisheries 6: 97–107.
14. De Schutter, O. 2011. The green rush: the global race for farmland and the rights of land users. Harv Int Law J 52: 503–559.
15. Driesen, DM. 2002. What's property got to do with it? Ecol Law Q 30: 1003.
16. 2009. A cautionary tale about ITQs in BC fisheries Ecotrust Issue Brief 8: 1–7.
17. Eggert, Håkan, and Mats Ulmestrand. 2007. Tenure rights and stewardship of marine resources: A co-managed Swedish shrimp fishery in a marine reserve. In 253rd ed. Göteborg University School of Business, Economics and Law Working Papers in Economics.
18. Eswaran, H, R Lal, and PF Reich. 2001. Land degradation: an overview. In Responses to Land Degradation, ed. Bridges EM, 20–35. Oxford Press, New Delhi.
19. 1995. Code of conduct for responsible fisheries. Food and Agriculture Organization (FAO), Rome.
20. 2004. Legislating for property rights in fisheries FAO Legislative Study 83: 3–153.
21. 2009. Black's Law Dictionary. West Publishing Company, St. Paul.
22. 2004. Individual fishing quotas: methods for community protection and new entry require periodic evaluation GAO-04-277. 1–51.
23. Hamilton, ND. 1998. The role of law in promoting sustainable agriculture: Reflections on ten years of experience in the United States. Drake J Agric Law 3: 423–431.
24. Hamilton, ND. 2010. Feeding our future: Six philosophical issues shaping agricultural law. Nebraska Law Rev 72: 210–257.

25. Hamilton, ND. 2011. Moving toward food democracy: Better food, new farmers and the myth of feeding the world. Drake Journal of Agricultural Law 16: 117–145.
26. Hardin, G, W. 1968. The tragedy of the commons. Science 162: 1243.
27. 1997. § 266.39.
28. Karp, JP. 1993. A private property duty of stewardship: Changing our land ethic. Environ Law 23: 735–762.
29. Kerr, S, J Sanchirico, and R Newell. 2003. Evaluating the New Zealand individual transferable quota market for fisheries management. Tradeable permits: Policy Evaluation, Design and Reform, Organisation for Economic Co-operation and Development. Social Science Electronic Publishing, Wellington.
30. Kooiman, J. 2005. Introduction. In Fish for Life: Interactive Governance for Fisheries, ed. Kooiman J, Bavinck M, jentoft S, Pullin R, 241–244. Amsterdam University Press, Amsterdam.
31. Kooiman, J, and S Jentoft. 2005. Hard choices and values. In Fish for life: Interactive governance for fisheries, ed. Kooiman, 285–302. Amsterdam University Press, Amsterdam.
32. Kooiman, J, M Bavinck, R Chuenpagdee, R Mahon, and R Pullin. 2008. Interactive governance and governability: An introduction. J Transdisciplinary Environ Stud 7: 1–11.
33. Kooiman, J, S Jentoft, M Bavinck, R Chuenpagdee, and R Sumaila. 2005. Meta-principles. In Fish for Life: Interactive Governance for Fisheries, ed. Kooiman J, Bavinck M, Jentoft S, Pullin R, 265–284. Amsterdam University Press, Amsterdam.
34. Kooiman, J, and S Jentoft. 2009. Meta-governance: Values, norms and principles, and the making of hard choices. Public Adm 87: 818–836.
35. Kurien, John. 2004. The blessing of commons: small scale fisheries, community property rights, and coastal natural assets. In 72nd ed, 23–54. University of Massachusetts (Amherst) Political Economy Research Institute Working Paper Series.
36. Leal, DR. 2000. Homesteading the oceans: the case for property rights in U.S. fisheries. PERC Policy Series Issue Number PS-19. 1–36.

37. LeVan, G. 1962. The usufructuary's obligation to preserve the property. Louisiana Law Rev 22: 808–818.
38. Ludwig, D, et al. 1993. Uncertainty, Resource Exploitation, and Conservation: Lessons from History. 17: 260.
39. Macinko, S. 2007. Fishing communities as special places: the promise and problems of place in contemporary fisheries management. Ocean Coastal Law J 13: 71–94.
40. Macinko, S, and DW Bromley. 2004. Property and fisheries for the twenty-first century: Seeking coherence. Vermont Law Review 28: 623.
41. Macinko, S, and S Schumann. 2008. The process of "property": Stasis and change in lobster management in southern New England. Vermont Law Review 33: 73.
42. Macinko, S, and W Whitmore. 2009. A New England dilemma: Thinking sectors through. Massachusetts Division of Marine Fisheries, Boston.
43. [http://catchshares.edf.org/sites/catchshares.edf.org/files/Japanese_Common_Fishing_Rights.pdf] McIlwain, Karly. 2013. Catch Shares in Action: Japanese Common Fishing Rights System. Environmental Defense Fund. .
44. Munro, GR, B Turris, C Clark, UR Sumaila, and M Bailey. 2009. Impacts of harvesting rights in Canadian pacific fisheries. Fisheries and Oceans Canada Statistics and Economic Analysis Series No. 1–3.
45. Myers, N, and C Raffensperger. 2006. Precautionary tools for reshaping environmental policy. MIT Press, Cambridge.
46. 2010. Toward sustainable agricultural systems in the 21st century. National Academies Press, Washington, DC.
47. 1999. Sharing the fish: toward a national policy on individual fishing quotas. National Academies Press, Washington, DC.
48. 1986. Iowa Ct. App.
49. Scott, A. 2000. Moving through the narrows: from open access to ITQs and self-government. Use of Property Rights in Fisheries Management (FAO Fisheries Technical Paper 404/1). Food and Agriculture Organization (FAO), Fremantle.
50. Squires, D. 1995. Individual transferable quotas as a fisheries management tool. Rev Fish Sci 3: 141–169.

51. Stevenson, GG. 1991. Common property economics: A general theory and land use applications. Cambridge University Press, Cambridge.
52. Sumaila, UR. 2010. A cautionary note on individual transferable quotas. Ecol Soc 15: 36.
53. Taylor, M. 1992. The economics and politics of property rights and common pool resources. J Nat Resour 32: 633.
54. 2012. 3:12-cv-00250-HR (2012) U.S.D.C. Alaska.
55. 1996. Poverty and hunger: Issues and options for food security in developing countries.
56. Ziff, B. 1996. Principles of property law. Carswell, Toronto.

Chapter 5

Conservation and Grazing in Australia's North-East: The Bridled Nailtail Wallaby

Fiachra Kearney[1], Ryan RJ McAllister[2], and, Neil D MacLeod[2]

[1]Bridled Nailtail Wallaby Trust, 61 Roderick Street, Ipswich, 4305, Australia

[2]CSIRO, Ecosystem Sciences, PO Box 2583, Brisbane, 4001, Australia

ABSTRACT

Australia's vast continent is dominated by semi-arid and arid landscapes that have been modified to support the development of an extensive livestock grazing industry. Historically, this development has come at great environmental cost, with wide-scale landscape degradation and loss of biodiversity, including small macropods. With the growing appreciation of environmental values and ecological services provided by grazing landscapes, the engagement of pastoral landholders is now central to contemporary conservation efforts. In this paper we explore the spluttering recovery of Australia's critically endangered bridled nailtail wallaby *Onychogalea fraenata*, once presumed extinct but now subject to a limited rehabilitation program in Queensland. We explore the 'fit' between management units and the scale of conservation challenges for the bridled nailtail wallaby, and then use this to frame the role of the private grazing industry in the governance of conservation actions. A centralised state conservation program has largely failed to stop the decline of the species, which remains critically endangered. We argue that non-state (privately) managed grazing properties working within a multi-level governance system that includes the state have a greater chance of conservation success because their actions can more appropriately match the scale of the problem at the implementation level. If the species recovers, the balance of management focus will need to shift towards broader scale actions such that localised disconnected sub-populations can successfully interbreed. By analysing the institutional failures that surround the bridled nailtail wallaby, we provide recommendations on how public institutions or policies can successfully catalyse private sector action at regional scales. These include avoiding economic incentives that may crowd out local stewardship, avoiding overly-authoritative state control (i.e. mono-centricity), and developing a multilevel governance structure that can strategically adapt its focus to the scale of various and shifting targets.

BACKGROUND

Attempts to protect global biodiversity are rife with examples of conservation failures for species for which sufficient scientific information actually exists to otherwise support successful conservation efforts (Millennium Ecosystem Assessment 2005). Economic pressure on landscape managers to maintain viable activities (especially food and fibre production to satisfy the needs of increasing human populations) and progressive land-use change are leading drivers of species' decline. Often though, these drivers are themselves symptoms of underlying institutional failures. Institutions are the social norms, and non-regulatory and regulatory rules that govern the actions that both drive conservation issues and prohibit conservation actions (see Young 2002). In this paper we explore the history of the conservation of the endangered bridled nailtail wallaby (*Onychogalea fraenata*). This species' once wide-scale semi-arid range has been heavily disturbed by predation from feral pests (e.g. foxes and dingoes) and extensive land clearing for pasture development for livestock grazing to the point that it is threatened by extinction. The sheer scale of habitat loss and predation load is such that, like many threatened savanna and woodland flora and fauna species, the wallaby population cannot feasibly be recovered and then protected within the limited network of formal conservation reserves that exist in the extensive grazing regions of Australia (e.g. McIntyre and Barrett 1992;Pressey 1995; Young and Howard 1995). Given that the larger part of the landscape is now privately owned and managed by pastoralists, it is inevitable that the solution to securing the wallaby's future will have to be found on that grazing land and the pastoralist community will potentially be key players in any serious efforts to recover the species. However conservation efforts for the wallaby remain fragmented and poorly coordinated and the broader pastoralist community, which derives little apparent value from the species presence on their land, is largely disconnected from both the species and its protection.

International evidence suggests that where communities are excluded from the governance of their local resources conservation efforts will generally fail. For example, in Africa's Maasailand poor relationships between the state and individual enterprises have been disempowering local communities thus hampering both conservation and livelihood outcomes (Homewood and Thompson 2009). In South America, international involvement in the efforts to protect the vicuna have worked perversely to both endanger and protect the species, but a lack of community engagement in active recovery efforts is ultimately threatening its sustainability (Lichtenstein and Carmanchahi this edition; McAllister et al. 2009a). At the other extreme, however, others warn that simply seeking to empower local resource users provides no guarantee for achieving conservation success (Bradshaw 2003). Questions have been raised over the credibility of the sustainability agendas of local communities and their capacity to actually implement them (Nelson and Agrawal 2008), as well as the credibility of efforts of central authorities to truly devolve power to local resource users (Bradshaw 2003).

A framework for exploring conservation problems is emerging around the problems of 'institutional fit' (see Cumming et al. 2006; Folke et al. 1998). This framework is centred on the concept that management that is required to put scientific information to good use will necessarily operate at temporal and spatial scales that are frequently inappropriate for the problem at hand. More broadly, the outcome of collective management will largely be determined by the suite of policy and private actors who 'interplay' around particular conservation issues. This collective construct of management is referred to as multilevel governance. The key to managing the problem of establishing appropriate institutional fit lies not merely in fitting the management to a particular scale, but rather exploiting the multilevel aspect of governance systems in order to adaptively coordinate the various conservation activities. In this context, pastoralists working at a local scale need to be given authority to undertake local conservation decisions, but with connectivity and accountability established within a broader system of governance.

Therefore, pastoralists holding the status of owners and managers of the greater proportion of local grazing landscapes need to be key partners in managing those landscapes for conservation outcomes. However, grazing per se is by no means the saviour of Australia's threatened biodiversity. The grazing industry working within the context of a long history of policy support has undertaken extensive land clearing and pasture development actions that have had a dramatic effect on landscape ecological values (McIntyre et al. 2002; Woinarski and Fisher 2003) and soil function (Kaur et al. 2007). High grazing intensity reduces cover and thereby increases sediment losses (Bartley et al. 2010); reduces total soil nitrogen and organic carbon (Beyer et al. 2011); interrupts natural fire cycles (e.g. Hodgkinson et al. 1984); and reduces the abundance of bird species (Martin et al. 2005) and small mammals (Woinarski and Fisher 2003).

Nevertheless, while substantial environmental degradation in Australia has been attributed to livestock grazing (NLWRA 2002; Tothill and Gillies 1993), the grazing industry is potentially a critical resource and a partner to national efforts to conserve biodiversity (Campbell 1995), and particularly some highly endangered species. There is no technical possibility of re-creating pre-European rangeland systems of Australia, nor would it be economically sensible to attempt to do so (Pannell 1999, 2000). These vast landscapes need to be actively managed to achieve either sustainable production or conservation outcomes, and policies of complete abandonment will offer limited benefit to any party. As such, working actively with the pastoral industry into the future is essentially the only realistic option open for improving biodiversity outcomes across large tracts of the landscape.

We focus on a northern Australian rangeland case-study, where a commercially-oriented livestock grazing land-use system has largely replaced one of traditional Aboriginal resource use. Specifically we recount conservation efforts of the bridled nailtail wallaby. It is important to note that for bridled nailtail conservation there are limited tangible co-benefits with grazing. Hence the problem is one of co-existence, and of how to structure a governance system specifically targeted for a multi-scale conservation problems that incorporates action in grazing enterprises. We argue the existence and effect of institutional inadequacies using scale mis-matches as our analytical framework. A key focus of the review is on how and why conversation of the species on non-state, privately managed grazing properties may have some success if managed appropriately and how state institutions might work to successfully upscale such non-state successes.

HISTORY OF THE DEMISE OF THE BRIDLED NAILTAIL WALLABY

The bridled nailtail wallaby is a small, solitary ash-grey macropod with striking markings across its shoulders resembling a white bridle, and a distinct black-stripe along its spine (Figures 1). Males and females weigh 5-8 kg and 4-6 kg respectively. of the three species in the genus, the northern nailtail wallaby (*O. unguifera)* is presently secure in status (Ingleby 1991), the crescent nailtail wallaby (*O. lunata*) is 'Presumed Extinct' (Burbidge 1983) and the bridled nailtail wallaby is classified as being 'Endangered' (Evans and Gordon 2008).

Figure 1: Bridled nailtail wallaby: (A) showing striped markings on back (© Project Kial), (B) with livestock on Avocet grazing enterprise (©Brian Siebel 2008) and, (C) monitoring (©Brian Siebel 2008).

The bridled nailtail wallaby was once distributed across millions of square kilometres ranging from Charters Towers in north Queensland down to the Murray River on the New South Wales-Victorian border. Habitat for the species varies through open grass woodlands (Eucalypt dominant) to dense shrublands (Acacia dominant), although it generally shows a marked preference for more fertile areas of the landscape and especially transition areas where dense shrubland and grassy woodlands meet (Evans and Gordon 2008; Gordon and Lawrie 1980). The animal appears to prefer to shelter in the thick shrubs and fallen timber by day and nocturnally graze shorter pastures, and when feeding in open pasture stays close to the edge of the thicker shelter habitat presumably to evade detection by predators (Evans and Gordon 2008; Tierney 1985).

Like many small-medium sized Australian mammals the populations of the bridled nailtail wallaby suffered precipitous declines following European colonisation and the expansion of pastoralism across the continent (Gordon 1983). In fact, there were no confirmed sightings of the wallaby between 1937 and 1973 and the animal was generally thought to have become extinct. There are many factors underlying the dramatic decline in wallaby numbers across its historic range, and these are thought to have been mutually reinforcing, such as significantly increased predation rates following the introduction of foxes and feral cats, extensive clearing of native vegetation (especially Acacia shrublands) for pasture development and the alteration of habitat structure and composition by further vegetation management and selective grazing by livestock where some of the original habitat remained (Lavery and Tierney 1985).

Grazing by sheep has been specifically implicated as a driver for the early phases of the decline in the wallaby's range because the animal relies heavily upon herbs and forbs for forage, with some selection of native grasses, and the species occupies a fairly narrow niche in the trophic system (Gordon and Lawrie 1980). Sheep grazing created direct competition for food and ultimately lead to a broad-scale compositional change within the shrubland and adjacent woodland understory. In the northern part of the wallaby's former range, beef cattle largely displaced sheep as the principal grazing animal and the competitive effect possibly came more through structural modification of the shrub understory through animal movement and trampling rather than direct competition for forage and consequent compositional change of the grass and forbs communities.

The major structural change in the northern habitat associated with expanding cattle numbers largely came through the wide scale clearing of the Acacia shrublands (especially Brigalow - *Acacia harpophylla)* and sowing of buffel grass (*Cenchrus ciliaris*) pastures under various publicly sponsored land development schemes from the early 1960's (QLAC 1968). Buffel grass typically alters the shrub understory structure and, if left unmanaged, can create dense physical barriers that extend over areas that are large enough to prevent the effective migration and immigration of smaller animals (Eyre et al. 2009; Smyth et al. 2009). This particular grass species can also out-compete native grasses and forbs, thereby decreasing food availability and selection for the bridled nailtail wallaby. Buffel grass can also support hotter fires than the maximum temperature that the acacia overstory can withstand, thus further removing suitable habitat for the wallabies (Miller et al. 2010).

A remnant population of the species was rediscovered on two grazing properties in Central Queensland in 1973 (Gordon and Lawrie 1980) in what is now Taunton National Park (Scientific) (see Figures 2E). This remnant population was immediately surveyed, the two properties on which it resided were acquired and a state (Queensland Government) conservation program was initiated by the National Parks Service. At the time of acquisition, the wallaby population was estimated at approximately 1400 individuals, with most of the animals located within the boundaries of Taunton National Park. From that time, however, there appears to have been a substantial decline in wallaby numbers despite state plans to protect the species.

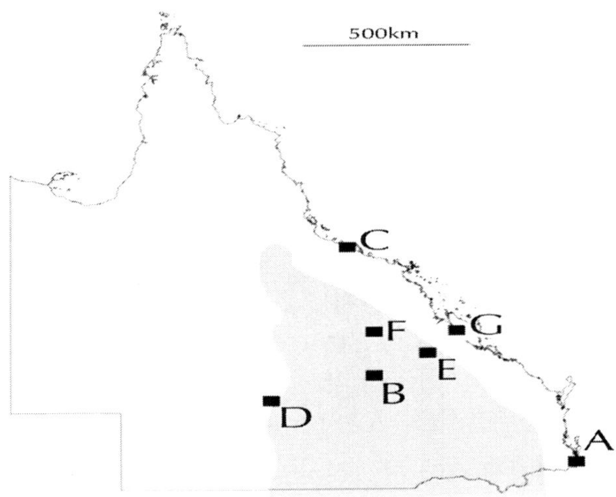

Figure 2: Key locations of bridled nailtail wallaby populations within Queensland: (A) Gold Coast; (B) Avocet Grazing; Property; (C) Townsville; (D) Idalia National Park; (E) Taunton National Park; (F) Gregory Mine; (G) private breeding facility. Shading indicates the approximate historical range.

MIS-MATCHES IN SCALE AS AN ANALYTICAL FRAMEWORK

The failure to conserve a species, and particularly to save it from extinction, can be seen simplistically as a failure in human behaviour. To positively influence species conservation efforts, human behaviours need to be understood in terms of the rules that govern them - social norms, and regulatory and non-regulatory rules that are applied by multiple scales of government. These rules are 'institutions' and can been categorised as either 'rules on papers' which refer to the literal interpretation of legislative rules about what actions can be enacted, or 'rules in use' which refer to how people will interpret the written rules and how they combine with social norms (Young 2002). Institutions also facilitate cooperation, and build future consequences into otherwise myopic decision making (Cumming et al. 2006).

Institutional settings have both spatial and temporal scales, and where conservation efforts do fail, it is often argued that the root cause of the failure lies in the scale of institutional settings not being aligned to the scale of the ecological processes that they purport to manage (e.g. Cumming et al.2006; du Toit 2009; Folke et al. 1998). We briefly consider the concept of mis-matches in scales and summarise their implications, before using this framework to explore the history of conservation efforts for the bridled nailtail wallaby.

Scale relates to the extent over which processes bear influence. We refer to the ecological spatial scale as the geographical extent of fundamental ecological processes (see Cumming et al. 2006). The sociological spatial scale refers to the extent of the collective influence of "rules, polices and informal and formal culture norms", where this collectivization occurs through levels of social and jurisdictional organisation. Sociological spatial scales therefore also refer to the geographical extent of influence (e.g. state boundaries, neighbourhood). Cumming et al. (2006) note that, there are both empirical and subjective measures of scale. Because the sociological scales involve composites of many actors and many institutions (e.g. Lubell et al. 2010) drawing scalar boundaries is necessarily more subjective. In fact the formation of many institutions is actually in directly response to the need to match ecological and management scales (e.g. see McAllister et al. 2009b).

Temporal scales are equally important in a framework for considering mis-matches of scale, and tend to be even more subjective as they relate to the memory or persistence of particular actions. For example, temporal scales may be driven by diverse factors such as the longevity of political decisions or the length a particular species' breeding cycle.

The crux of the framework is that where the scale of the ecological processes is not aligned to the scale over which management is attempted, effective conservation management is unlikely to emerge and poor outcomes are most likely to prevail. Two particular mis-matches of scales are common when attempting to address conservation problems. The first mis-match of scale involves politically driven sociological time scales that are much shorter than the time commitment required to successfully fulfil a particular conservation action (Table 1). The second mis-match of scale concerns ecological problems occurring at broad spatial scales that are addressed by management actions that are applied well below the required spatial scale (or configuration).

Table 1: Implications of scalar mis-matches, adapted from Cumming et al. (2006)

Scalar mis-matches		Implications
Spatial scales		
broad sociological	fine ecological	Too many managers. Micromanager syndrome; lack of local knowledge leads to poor adaptability
fine sociological	broad ecological	No solutions for global problems, inconsistency and competition across regions
Temporal scales		
long sociological	short ecological	Response times to slow
short sociological	long ecological	Management decisions lack continuity and consistency, and limited consideration for the future

Kearney *et al.*

Kearney *et al. Pastoralism: Research, Policy and Practice* 2012 **2**:20, doi: 10.1186/2041-7136-2-20.

How these mis-matches of scales will actually play out in practice relates to the concept of governance. Within any conservation agenda, there inevitably exist multiple management actions occurring at various scales that are the responsibilities of multiple actors whom likewise operate at multiple scales. Governance can take many forms.

For example, power can be vested entirely with a centralised authority (i.e. mono-centric) or be distributed between public and private actors operating across a range of scales (i.e. multi-level), and each approach has advantages and disadvantages (see Termeer et al. 2010). When applied to the context of addressing conservation problems, it is necessary to think about what governance arrangement may work best in the context of having a set of required conservation actions to produce desired outcomes. It is also necessary to consider what governance arrangements will facilitate the best level of coordination across those necessary actions. This coordinating role can also take various forms. Traditionally, under the mono-centric governance model the central authority typically undertakes all coordination activities. Under the multi-level governance model, the coordination role adaptively emerges through the interplay between various levels of governance and across various scales (see Cash et al. 2006; Lubell et al. 2010).

Our purpose here, however, is not to specifically interrogate various models of governance, but to use this as a framework to consider the potential role of pastoralists as local land management agents for mitigating problems of institutional fit within these broader governance systems.

MONO-CENTRIC CONSERVATION OF THE BRIDLED NAILTAIL WALLABY

In Queensland, the focus of this case example, all native fauna and flora are the exclusive property of the state (i.e. owned on behalf of the public by the Government of Queensland), including the bridled nailtail wallaby on both public and privately owned land. A state-based conservation program has been operating for the wallaby in Queensland from the time that the remnant population was re-discovered in 1973, During that time a some ecological, biological and management research has been undertaken relating to the species (e.g. Davidson 1991; Ellis et al. 1992; EPA/QPWS 1998;Fisher 1998; Gordon and Lawrie 1980; Horsup and Evans 1992; Lundie-Jenkins and Lowry 2005). Many management recommendations have been developed for conservation of the bridled nailtail wallaby from this body of knowledge, and some have been implemented. For example, feral cats are recognised to be major predators of young wallabies and thought to be responsible for suppressing juvenile recruitment, and dingoes are held responsible for the majority of adult deaths (Lundie-Jenkins and Lowry 2005). Therefore, effective predator control is considered to be essential for conservation success of the species and has been a recurring recommendation although only limited resources have actually been made available by the state to carry it out.

A captive-breeding population of wallabies was established in 1991 at a state run facility in north Queensland. In 1992 and 1997, two additional state controlled captive-breeding centres were established on the Gold Coast and at Gregory Mine, and also a private zoo facility at Rockhampton (Figures 2). Some animals were also translocated from Taunton National Park with the aim of establishing two additional discrete populations, which is a standard security measure for highly endangered species. The translocation sites were located at Idalia National Park (commenced 1996) and 'Avocet' which is 5,500-ha privately owned and operated cattle grazing property located near Springsure in Central Queensland (commenced 2001) (MacLeod et al. 2010). Both translocation sites were considered to meet the necessary habitat requirements for the bridled nailtail wallaby and also offer good scope for the implementation of necessary management procedures. A captive-breeding facility was also established at the Idalia National Park site, albeit temporary in nature.

Substantial financial resources have been expended on the state recovery program for the bridled nailtail wallaby -AUD $3.39 million between 2005 and 2009, equating to approximately US$ 2.7 million in 2009 - (Lundie-Jenkins and Lowry 2005), including land acquisition, human and other resources, education and protected area management. In addition, the state has invested in research into the ecology, behaviour and conservation needs of the species. The stated operational goal of the responsible agency is "to significantly improve the conservation status of the bridled nailtail wallaby.." (4th edition of the recovery plan for the bridled nailtail wallaby, 2005-2009). There were 8 specific targets to be reached which include (1) to enhance or maintain existing free-range populations of bridled nailtail wallabies at Taunton and Idalia National Parks, and Avocet Nature Refuge and (2) to establish two or more additional populations of the species (>100 individuals) within Queensland. A significant share of the conservation resources were directed to Taunton National Park which holds the source population for all recovery efforts. The Taunton population is also the most critical genetically (Sigg 2006) because it contains the highest allelic diversity (Seddon 2008). Pople et al. (2001) considered the translocation of animals to Idalia to be a success because the population there was estimated to have increased from 133 founders to in excess of 400 individuals. However, all captive breeding programs are expensive requiring challenging and intensive management to maintain genetic diversity (Stockwell et al. 1996) and to ensure that the translocated populations are not overly compromised. Sigg (2006) has subsequently reported progressive genetic loss within the translocated population in Idalia relative to the source population at Taunton National Park.

The estimated Taunton wallaby population declined markedly between 1991and 1994, from approximately 1400 individuals to around 450 individuals (Pople et al. 2001), the rapid decline being blamed primarily on folder deficiencies relating to severe drought conditions that prevailed in the region (Clancy and Porter 1993). In 2008 the Taunton population was estimated to have declined further to only 75 individuals (SE 63-93, *95%*) (Taunton National Park, unpublished workshop report). The breeding facility at Gregory Mine was also closed due to nutritional degradation of the natural foods on-site during a prolonged drought (Lundie-Jenkins and Lowry 2005).

From the late 1990s the NGO, Australian Wildlife Conservancy, has conducted a bridled nailtail wallaby breeding program on a privately owned, managed and fully enclosed and predator free facility known as 'Scotia Sanctuary', in southwestern New South Wales. As of the end of 2011 the sanctuary is reported to be holding in excess of 1500+ individual animals (AWC 2011). While this particular captive-breeding effort appears to have been successful in terms of the numbers of individual animals produced, the resident population is yet to be exposed to natural predation levels and appears to be confronted with substantial genetic losses in many of the individual animals (Seddon 2008). A very limited number of these animals may be able to be utilised to augment other populations if the necessary state approval can be obtained, but the core of the wallaby conservation efforts and all remnant and wild translocated populations presently remain in Queensland.

In 2011 the bridled nailtail wallaby still remains on Queensland, Australian and international endangered species lists, no new populations have been established, and the existing populations appear to still be declining. It would seem fair to conclude at this stage that the present mono-centric conservation approach that has been adopted for the bridled nailtail wallaby has not been particularly successful. An alternative approach would seem to warrant further scrutiny.

MULTI-LEVEL CONSERVATION ON PRIVATELY MANAGED GRAZING LANDS

From 2001, the owners of the 5,500 ha grazing property 'Avocet' have worked with the Queensland Government (Parks and Wildlife Service) to establish a quasi-state managed population of bridled nailtail wallabies on a formal reserve of approximately 1,100 ha (see previous section). This action has involved considerable costs being borne by private individuals having continued to forgo the full use of the habitat reserve land which is potentially productive grazing land, and incurring ongoing land management costs associated with maintaining the reserve (e.g. fire breaks, fencing). Several non-state initiatives have also worked with Avocet. An organisation of independent and voluntary licensed shooters (Conservation and Wildlife Management Queensland) was established in order to counter feral animal predation on native endangered species. This utilises trained individual volunteer shooters using their own equipment to control feral cats, foxes and dogs on conservation reserves. A small publicly registered conservation trust dedicated to bridled nailtail wallaby conservation was also founded (Bridled Nailtail Wallaby Trust) to facilitate and apply research, coordinate conservation efforts of non-state actors, lobby the state and draw attention to the failings of conservation. Both organisations approached the owner of 'Avocet' to assist with the management of the species, and a multi-level conservation approach was subsequently formed to achieve that goal.

The 'Avocet' private grazing property is presently the only location in Queensland where appropriate species and habitat management practices can be actively applied by non-state actors without being subject to the significant constraints that are presently imposed on non-government organisations operating on state controlled land. The private land-NGO partnership offers much greater flexibility and the opportunity to act pro-actively in the species conservation. 'Avocet's owner/manager, with support from NGOs, has initiated a series of conservation management activities at a level of effort that has not been previously employed. 'Avocet' is surrounded by other grazing enterprises, and hence there is unrealised potential to engage with surrounding pastoralists to address coordination of management conservation. While we hypothesise below that this governance model has the greatest potential to lead to sustained conservation outcomes, success has not yet been achieved.

A research study conducted on 'Avocet' in 2008 found approximately 40 individual animals (Kingsley 2009), which was far lower than expected given the progressive release of 166 animals from 2001. The reason for the apparent lack of success of the translocation effort on 'Avocet' remains unclear because there have not been adequate post-release surveys conducted, but a range of biotic and abiotic factors are likely to have had an impact on the bridled nailtail wallaby population since translocation activities commenced. As a consequence, a major focus on 'Avocet' has been directed to developing institutional capacity and structure in a way that may promote an effective conservation outcome in the future – viz. effective population monitoring, application of research recommendations, immediate response to disease or fire emergencies, formal adaptive management actions that are actually implemented.

Local conservation actions on Australian privately managed grazing enterprises are critical, but from both a pragmatic or a legal perspective they alone is unlikely conserve the bridled nailtail wallaby, even with support from NGOs. The NGOs that may partner private grazing enterprises are small and have a limited private support base, and also operate in a congested space in terms of public support for conservation initiatives (e.g. Greening Australia, Bush Heritage, Koala Foundation appeals etc.). Funding is typically a constant challenge for conservation projects that are focussed on relatively little-known species (Seddon et al. 2005). Hence for aspects of threatened species management that are costly and require highly specialised skills, such as genetic diversity management, the role of state is also critical. However, arranging and then managing state-NGO partnerships can be challenging. Public lobbying on behalf of either the species or its supporting NGO can create tension, and no formal mechanisms exist to cooperate on conservation efforts. Nonetheless, as private grazing land offers considerable potential to facilitate the implementation of effective conservation programs it is well worth proceeding with or without state cooperation insofar as this is possible.

Potential Win-Wins

While generally we consider that conservation of the bridled nailtail wallaby can co-existence with grazing, buffel grass management provides at least one example where pastoralist management can assist in meeting both conservation and grazing goals. Buffel grass is a widely planted pasture species throughout the grazing lands of central and western Queensland, is highly invasive, and is considered a major pest species by many conservation bodies (Miller et al. 2010). However, it is also an important fodder species for beef cattle production (Humphries 1967). Now found across vast tracts of northern and central Australia, its control on conservation lands is highly problematic, and it has been implicated in the ongoing decline of the bridled nailtail wallaby population on Taunton National Park (Taunton National Park, unpublished workshop report).

Cattle grazing, especially using intensive periodic short duration grazing cycles is accepted as an important management tool for controlling buffel grass (Butler and Fairfax 2008). While further buffel introductions may severely hinder conservation efforts, cattle grazing is important management tools where this and other introduced grass species already exist, and working with pastoralists sympathetic to conservation goals on their property is imperative.

Conservation Ethos

While profit is the underpinning motivator of many pastoralists' behaviour (Richards et al. 2005), evidence suggests that most pastoralists also have a strong environmental stewardship ethic (e.g. Green and MacLeod 2002; Greiner and Gregg 2011). This ethos needs support to achieve the required conservation goals. Yet there is potential for overly centralised policy to smother this stewardship by disempowering and ultimately de-motivating local stakeholders, and inducing complacency from local pastoralists concerning species conservation efforts on their holdings. Furthermore, insights from the emerging discipline of experimental economics suggest that attempting to monetarise altruistic behaviours, such as environmental stewardship, risks crowding out such actions (see Reeson and Tisdell 2008). This point to some value in the devolution of decision-making power to local stakeholders where this can encourage pastoralists existing stewardship motivations.

DISCUSSION

Governing from the Bottom-up

Our framework for understanding conservation failures for the endangered bridled nailtail wallaby has focussed on an examination of institutional failures in the dominant paradigm as well as advancing alternatives that offer greater potential success. A critical question that has been raised is whether or not bridled nailtail wallaby conservation management issues and resultant actions are necessarily being targeted at a scale that is appropriate to effectively address the problem. We are particularly interested in the potential role of pastoralists in effective landscape resource conservation, including local protection of endangered species, and so the framework seeks to understand their role within the broader system of governance. It is important to note, however, that such an institutional framework does not assume that there is a single panacea to solve all conservation problems. Regardless of the purpose, institutions are rarely, if ever, perfect.

Institutions that are created to address a specific purpose are not merely the product of a single issue. Institutions necessarily respond to geo-political contexts, ecological and economic drivers, path dependencies and also to forces from other competing and complementary institutions. Complexity can foster resilience, but not perfection. Furthermore, conservation issues and drivers change continually, so a perfect institutional solution will only be "perfect" for particular moments in time and space. Therefore, in the absence of finding permanently perfect institutions to resolve complex conservation problems, a strong case exists for consciously building adaptive capacity into the governance of such problems.

Single conservation issues generally involve a range of management problems that individually operate across many scales (Table 2). For the bridled nailtail wallaby, captive breeding and on-property landscape restoration can maintain localised and isolated sub-populations, which is spatially a small-scaled problem. However, to truly recover the species across a significant range is a larger-scale problem that requires genetic interplay between the recovering sub-populations. Prior to European settlement, this genetic interaction was naturally maintained through species mobility across an unfragmented landscape. This is no longer possible, of course, and now requires conscious and intensive management.

Table 2: Major conservation sub-tasks and associated scales

Sub tasks	Spatial scale required	Temporal scale required	Locations
1. Captive breeding	<1 ha	Years	
2. Isolated natural habit populations	400-600 ha	Decades	'Avocet' – operational grazing property
3. Genetic interplay between populations	Connections between the system parts	Decades	Queensland
4. Connected natural habit populations	>500,000 ha	Centuries	Nowhere - unlikely

Kearney *et al.*

Kearney *et al. Pastoralism: Research, Policy and Practice* 2012 **2**:20, doi: 10.1186/2041-7136-2-20

Under a mono-centric governance model, species management may involve large sociological scales, yet is typically driven by political agendas drawn on relatively short temporal scales. Effective conservation management is necessarily built around much longer time horizons. In the specific case of conservation needs of the bridled nailtail wallaby, institutional coordination could realistically be centred around the state action plan for the species, with each actor/institution agreeing on clear responsibility for certain achievements. However, our governance-based framework would suggest that a lack management continuity and consistency, and limited consideration for the future are likely to be encountered as major problems (Table 1). Across the entire conservation agenda, some spatial mis-matches seem probable given various actions have differing requirements in terms to the scale of management (Table 2).

There is a clear need for broad sociological scaled management that can efficiently coordinate the genetic interplay across sub-populations, including those located on disparate privately owned and managed grazing properties and in captive breeding sites. Yet, if a broad sociological scaled institution were to manage all aspects of bridled nailtail wallaby conservation, it will likely be too broad to promote effective local action based on adaptive management principles - local conservation actions would likely be ineffectively micro-managed from above and based on decisions that lack local knowledge and the ability to take short-term actions as dictated by evolving local circumstances. In sum, such management lack adaptability (Table 1) and are unlikely to be effective. A mono-centric form of conservation governance is troubling in this context because even where conservation actions are locally driven, the centralisation of power can effectively scale all actions to that of the state-based, broad sociological scaled management.

This can be clearly illustrated by cases in which local scale conservation actions are having some success. 'Avocet' demonstrates how conservation efforts on privately owned and managed grazing properties can work at the implementation level because at that scale the local managers have relatively more decision-making autonomy than is the case for taking local actions on state managed land, and has external support to which the landholder also has direct input. Private landholder-NGO partnerships can help to support conservation efforts because they operate at the scale where direct action is required. Importantly, operation at local scales requires more than simply local managers having local issues as their focus. Private landholder-NGO partnerships potentially have the ability to make decisions at local scales because they operate in a decision-making environment that is relatively free from the inertia of bureaucracy. Such arrangements, therefore, have a capacity to make and apply sound decisions in a timely manner, and the absorption of land management costs by the participating pastoralists is part of their daily business. The private land-NGO partnership basically offers considerably more flexibility and the opportunity to act proactively.

The challenge then is how to get the best of both worlds? That is, how do you allow localised conservation to remain unhindered by slower broader scale management, while at the same time coordinating local scale management activities in order to ensure that the ecological function of a dispersed population can be re-created.

In this instance, by analysing the institutional failures, and showing how the private sector can be an effective conservator, we provide recommendations on how public institutions or policies can catalyse private sector action at appropriate regional scales. The case of the bridled nailtail wallaby provides a few clear messages for species conservation that may be generalised.

Strong centralised state control is unlikely to be the answer for effective conservation management because this potentially breaks the necessary match of scales between local land management action and localised conservation needs and outcomes. Rather, the broad aim for effective governance should establish structures that will foster local conservation management action and autonomy, and to place local actors within a broader multi-level governance system still requires accountability (but not necessarily in a hierarchical sense).

Economic incentives are an important consideration in terms of creating local action by pastoralists, especially where desirable conservation outcomes carry significant elements of the provision of public goods (Willis 1997), but caution is needed in designing such reward systems. Pastoralists already have a strong stewardship ethic (Greiner and Gregg 2011) and this class of stakeholders is not disinterested in environmental outcomes that flow from their decision making actions that are largely centred on production issues (e.g. Green and MacLeod 2002). Economic incentives, while appropriate to addressing externalities involving under-provision of environmental services such as biodiversity conservation (Pannell 2008), need to be careful not to crowd out pastoralists' intrinsic conservation ethics (Greiner and Gregg 2011) and altruistic behaviours (see Reeson and Tisdell 2008). Hence rather than focusing excessively on the need to create payments for conservation, the aim should be on harnessing increased commitment of pastoralists to undertaking additional conservation efforts by mitigating the costs of doing so, and to formally acknowledge their already significant contribution to land resource stewardship (Curtis and Delacey 1998).

The most challenging issue for implanting this model of governance is related to organising the broad sociological management, particularly in terms of co-ordinating the roles and actions of the prospective partners. If we accept the argument that a mono-centric governance model is unlikely to promote effective conservation actions for endangered species such as the bridled nailtail wallaby, then some form of decentralised coordination is required to ensure that the dynamic interactions between multiple actors across multiple forums does lead to good conservation outcomes.

Policy decisions and plans do not always have their intended impact, nor should they realistically be expected to. The reason is that 'impact' is rarely a function of a single policy, but instead is determined by a system of multiple, overlapping institutions. The interplay between the multiple institutions is largely determined by the ways in which various policy actors choose to engage with each other and the target issues across these multiple institutions. Further, the motivation for policy actors to engage in particular institutions is commonly one of self-interest (Lubell et al. 2010). With no unifying master plan, while an emergent coordinated approach to policy can be managed, it cannot be strictly controlled.

Regardless of the context, the influence of an individual actor is largely determined by the extent to which their participation in a given issue modifies the dynamic actor-institutional networks that emerge around such settings. The structure of these policy networks, that support bridled nailtail wallaby conservation, need greater attention. There is no single correct solution to this issue, and policy networks can range from top-down to bottom-up and designed or self-organized (see Hahn 2011). The important consideration for effectiveness is that networks that are established combine the requirements of accountability, and legitimacy of various responsibilities.

CONCLUSIONS

Given the large scales over which the landscapes of the Australian continent have been modified to support an extensive grazing industry, in contrast to the limited scale of the public conservation reserve system, it is clear that the future effectiveness of resource and nature conservation efforts must necessarily lie with the application of environmental stewardship at the local scale. Successive Australian governments have sought to devolve natural resource management powers to local communities (e.g. Landcare), but many questions surround the legitimacy of devolution and whether it is just an alternate form of centralised intervention (Lane et al. 2009; Marshall 2009). Nevertheless, adequate devolution of responsibility and decision-making powers is essential if vulnerable species and habitats are to be adequately conserved. The state-based conservation program for the bridled nailtail wallaby has largely failed to date to increase the population size, distribution and genetic diversity of the species, despite some individual successes. An examination of the scales of management suggests that for conservation outcomes on privately managed grazing enterprises to be positive, a new *modus operandi* for conservation governance is urgently needed. This new model should be one that enables effective multi-level coordination of actors and actions and reduces the downside limits of state mono-centricity. Such *modus operandi* are taking shape. There is no strong evidence as yet that such governance is already leading to successful on ground conservation outcomes. Yet without new models the risk is strong that a wide community of otherwise potentially willing pastoralists will be de-motivated from undertaking serious conservation efforts on their holdings, the area of land under effective conservation management will ultimately be reduced, and personal accountability for failure will remain weak. As rangeland environmental values are increasingly recognised outside of the regions in which they are generated, increased accountability of local stewardship actions will be required. This needs to be balanced with a genuinely inclusive governance structure in which the various aspects of complex conservation problems can be managed concurrently at appropriate scales.

AUTHORS' CONTRIBUTIONS

FK, RRJM and NDM co-authored the manuscript with respective foci on landscape ecology, governance and institutions, and grazing enterprises. All authors read and approved the final version of the manuscript.

ACKNOWLEDGEMENTS

Many thanks to Hugo Spooner the owner of 'Avocet', the collective membership of Conservation and Wildlife Management (Qld) and the individuals within the Queensland State Government for their commitment to the conservation of the bridled nailtail wallaby. RRJM acknowledges feedback from participants of the ARC Centre of Excellence for Environmental Decisions Workshop 22-26 August. Thanks to Bruce Taylor and Angela Guerrero Gonzalez for helpful comments on the manuscript.

REFERENCES

1. AWC. 2011. Scotia: new horizons for the Bridled Nailtail Wallaby. Australian Wildlife Conservancy. Wildlife Matters Summer 2011 (12): 12–13.
2. Bartley, R, JP Corfield, BN Abbott, AA Hawdon, SN Wilkinson, and B Nelson. 2010. Impacts of improved grazing land management on sediment yields, Part 1: Hills lope processes. Journal of Hydrology 389: 237–248.
3. Beyer, S, A Kinnear, LB Hutley, K McGuinness, and K Gibb. 2011. Assessing the relationship between fire and grazing on soil characteristics and mite communities in a semi-arid savanna of northern Australia. Pedobiologia 54: 195–200.
4. Bradshaw, B. 2003. Questioning the credibility and capacity of community-based resource management. Canadian Geographer-Geographe Canadien 47: 137–150.
5. Burbidge, AA. 1983. Crescent Nailtail Wallaby Onychogalea lunata. In The Australian Museum Complete Book of Australian Mammals, ed. Strahan R, Angus and Robertson, Sydney.

6. Butler, DW, and RJ Fairfax. 2008. Buffel Grass and fire in a Gidgee and Brigalow woodland: A case study from central Queensland. Ecological Management and Restoration 4: 120–125.
7. Campbell, CA. 1995. Landcare: Participative Australian approaches to inquiry and learning for sustainability. Journal of Soil and Water Conservation 50: 125–131.
8. Cash, DW, WN Adger, F Berkes, P Garden, L Lebel, P Olsson, L Pritchard, and O Young. 2006. Scale and cross-scale dynamics: Governance and information in a multilevel world. Ecology and Society 11: 12.
9. Clancy, TF, and G Porter. 1993. Bridled nailtail wallaby (Onychogalea fraenata) recovery plan progress report No. 1. EPA/QPWS Conservation Strategy Branch. Unpublished report to ANPWS, June 1993.
10. Cumming, GS, DHM Cumming, and CL Redman. 2006. Scale mismatches in social-ecological systems: causes, consequences, and solutions. Ecology and Society 11: 14.
11. Curtis, A, and T Delacey. 1998. Landcare, stewardship and sustainable agriculture in Australia. Environmental Values 7: 59–78.
12. Davidson, C. 1991. Recovery plan for the bridled nailtail wallaby (Onychogalea fraenata). Unpublished report to ANPWS, ANPWS Endangered Species Program. December 1991
13. du Toit, JT. 2009. Addressing the mismatches between livestock production and wildlife conservation across spatio-temporal scales and institutional levels. In Wild rangelands: conserving wildlife while maintaining livestock in semi-arid, ed. Toit JT, Koch R, Deutsch JC, Chichester, John Wiley & Sons.
14. Ellis, BA, PJ Tierney, and TJ Dawson. 1992. The diet of the bridled nailtail wallaby (Onychogalea fraenata): site and seasonal Influences and dietary overlap with the black-striped wallaby (Macropus dorsalis) and domestic cattle. Wildlife Research 19: 65–78.
15. EPA/QPWS. 1998. Taunton National Park (Scientific) (Brigalow Belt biogeographic region): management plan. Environmental Protection Agency, Brisbane.

16. Evans, M, and G Gordon. 2008. Bridled Nailtail Wallaby. In The Mammals of Australia, ed. Dyck S, Strahan R, New Holland, Sydney.
17. Eyre, TJ, J Wang, MF Venz, C Chilcott, and G Whish. 2009. Buffel grass in Queensland's semi-arid woodlands: response to local and landscape scale variables, and relationship with grass, forb and reptile species. Rangeland Journal 31: 293–305.
18. Fisher, D. 1998. Behavioural ecology and demography of the bridled nailtail wallaby, Onychogalea fraenata. University of Queensland PhD Thesis, Brisbane.
19. Folke, C, L Pritchard, F Berkes, J Colding, and U Svedin. 1998. The problem of fit between ecosystems and institutions. International Human Dimensions Programme on Global Environmental Change (IHDP), Bonn.
20. Gordon, G, and BC Lawrie. 1980. The rediscovery of the bridled nailtail wallaby Onychogalea fraenata, (Gould) (Marsupialia: Macropodidae), in Queensland. Australian Wildlife Research 7: 339–345.
21. Green, JL, and ND MacLeod. 2002. Barriers and opportunities for adoption. In Managing and conserving grassy woodlands, ed. McIntyre S, McIvor JG, Heard KM, 177–200. CSIRO Publishing, Melbourne.
22. Greiner, R, and D Gregg. 2011. Farmers' intrinsic motivations, barriers to the adoption of conservation practices and effectiveness of policy instruments: Empirical evidence from northern Australia. Land Use Policy 28: 257–265.
23. Hahn, T. 2011. Self-organized governance networks for ecosystem management: who is accountable? Ecology and Society 16: 18.
24. Hodgkinson, KC, N HG, GE Griffin, JC Noble, and MD Young. 1984. Management of vegetation with fire. In Management of Australia's Rangelands, ed. Harrington GN, Wilson AD, Young MD, CSIRO Publishing, Melbourne.
25. Homewood, K, and M Thompson. 2009. Social and economic challenges for conservation in east African rangelands: land use, livelihoods and wildlife change in maasailand. In Wild rangelands: conserving wildlife while maintaining livestock in semi-arid, ed. Toit JT, Koch R, Deutsch JC, Chichester, John Wiley & Sons.

26. Horsup, A, and M Evans. 1992. Predation by Feral Cats, Felis catus, on an endangered marsupial, the bridled nailtail wallaby, Onychogalea fraenata. Australian Mammalogy 16: 85–86.
27. Humphries, LR. 1967. Buffel grass (Cenchrus cilaris) in Australia. Tropical Grasslands 1: 123–124.
28. Ingleby, S. 1991. Distribution and status of the Northern Nailtail Wallaby, Onychogalea unguifera (Gould 1841). Wildlife Research 18: 655–676.
29. Kaur, K, RK Jalota, and DJ Midmore. 2007. Impact of tree clearing on soil attributes for a pastoral property in central Queensland, Australia. Soil Science 172: 516–533.
30. Kingsley, L. 2009. Evaluation of the Success of an Off-park Translocation of the Endangered Bridled Nailtail Wallaby, Onychogalea fraenata. University of Queensland MSc Thesis, Brisbane.
31. Lane, MB, A Haygreen, TH Morrision, and J Woodlands. 2009. Will regionalisation achieve intergated natural resource management? In Contested Country, ed. Lane MB, Robinson CJ, Taylor B, 59–74. CSIRO, Melbourne.
32. Lavery, HJ, and PJ Tierney. 1985. Scarcity and Extinction. In The Kangaroo Keepers, ed. Lavery HJ, University of Queensland Press, Brisbane.
33. Lichtenstein, G, and P Carmanchahi. this edition. Payún Matrú: an alternative model for guanaco sustainable use. Pastoralism this edition
34. Lubell, M, AD Henry, and M McCoy. 2010. Collaborative Institutions in an Ecology of Games. American Journal of Political Science 54: 287–300.
35. Lundie-Jenkins, G, and J Lowry. 2005. Recovery plan for the bridled nailtail wallaby (Onychogalea fraenata) 2005-2009. Report to the Department of Environment and Heritage (DEH). Environmental Protection Agency/Queensland Parks and Wildlife Service, Brisbane.
36. MacLeod, ND, FC Kearney, H Spooner, and M Woods. 2010. Conservation of critically endangered wildlife species - scope for private landholder and citizens' action. 16th Biennial Conference Papers. Australian Rangeland Society, Burke.

37. Marshall, GR. 2009. Can community-based NRM work at the scale of lagre regions? Exploring the roles of besting and subsidiarity. In Contested Country, ed. Lane MB, Robinson CJ, Taylor B, 43–57. CSIRO, Melbourne.
38. Martin, TG, PM Kuhnert, K Mengersen, and HP Possingham. 2005. The power of expert opinion in ecological models using Bayesian methods: Impact of grazing on birds. Ecological Applications 15: 266–280.
39. McAllister, RRJ, D McNeill, and IJ Gordon. 2009. Legalizing markets and the consequences for poaching of wildlife species: the vicuna as a case study. Journal of Environmental Management 90: 120–130.
40. McAllister, RRJ, DM Stafford Smith, CJ Stokes, and F Walsh. 2009. Patterns of accessing variable resources across time and space: desert plants, animals and people. Journal of Arid Environments 73: 338–346.
41. McIntyre, S, and GW Barrett. 1992. Habitat variegation, an alternative to fragmentation. Conservation Biology 6: 146–147.
42. McIntyre, S, JG McIvor, and KM Heard. 2002. Managing and conserving grassy woodlands. CSIRO Publishing, Melbourne.
43. Millennium Ecosystem Assessment. 2005. Ecosystem and human well-being. Island Press, Washington DC.
44. Miller, G, M Friedel, P Adam, and V Chewings. 2010. Ecological impacts of buffel grass (Cenchrus ciliaris L.) invasion in Central Australia - Does field evidence support a fire invasion feedback? Rangeland Journal 32: 353–365.
45. Nelson, F, and A Agrawal. 2008. Patronage or participation? Community-based natural resource reform in sub-saharan Africa. Development and Change 39: 557–585.
46. NLWRA. 2002. Landscape health in Australia: A rapid assessment of the relative condition of Australia's bioregions and sub-regions. Environment Australia and the National Land and Water Resources Audit, Canberra.
47. Pannell, DJ. 1999. Explaining non-adoption of practices to prevent salinity in Western Australia: Implications for policy. Agricultural and Resource Economics, University of Western Australia, Nedlands.

48. Pannell, DJ. 2000. Salinity policy: A tale of fallacies, misconceptions and hidden assumptions. Department of Agricultural and Resource Economics, University of Western Australia, Nedlands.
49. Pannell, DJ. 2008. Public benefits, private benefits, and policy intervention for land-use change for environmental benefits. Land Economics 84: 225–240.
50. Pople, AR, J Lowry, G Lundie-Jenkins, TF Clancy, HI McCallum, D Sigg, D Hoolihan, and S Hamilton. 2001. Demography of bridled nailtail wallabies translocated to the edge of their former range from captive and wild stock. Biological Conservation 102: 285–299.
51. Pressey, B. 1995. Conservation reserves in NSW: Crown Jewels or leftovers? Search 26: 47–51.
52. QLAC. 1968. The brigalow story. Land Administration Commission, Department of Lands, Brisbane.
53. Reeson, AF, and JG Tisdell. 2008. Institutions, motivations and public goods: An experimental test of motivational crowding. Journal of Economic Behavior & Organization 68: 273–281.
54. Richards, C, GA Lawrence, and N Kelly. 2005. Beef production and the environment: Is it really 'hard to be green when you are in the red'? Rural Society Journal 15: 192–209.
55. Seddon, J. 2008. Genetic analysis of Bridled Nailtail Wallabies for Captive Breeding Programme. Unpublished report, School of Veterinary Science, University of Queensland.
56. Seddon, PJ, PS Soorae, and F Launay. 2005. Taxonomic bias in reintroduction projects. Animal Conservation 8: 51–58.
57. Sigg, DP. 2006. Reduced genetic diversity and significant genetic differentiation after translocation: Comparison of the remnant and translocated populations of bridled nailtail wallabies (Onychogalea fraenata). Conservation Genetics 7: 577–589.
58. Smyth, A, M Friedel, and C O'Malley. 2009. The influence of buffel grass (Cenchrus ciliaris) on biodiversity in an arid Australian landscape. Rangeland Journal 31: 307–320.
59. Stockwell, CA, M Mulvey, and GL Vinyard. 1996. Translocations and the preservation of allelic diversity. Conservation Biology 10: 1133–1141.

60. Termeer, CJAM, A Dewulf, and M van Lieshout. 2010. Disentangling Scale Approaches in Governance Research: Comparing Monocentric, Multilevel, and Adaptive Governance. Ecology and Society 15.
61. Tierney, PJ. 1985. Habitat and ecology of the bridled nailtail wallaby with implications for management. M. Sc. Thesis. Queensland Institute of Technology, Brisbane.
62. Tothill, JC, and C Gillies. 1993. The pasture lands of northern Australia. Tropical Grasslands Society of Australia, Brisbane. Occasional Publication No.5.
63. Willis, IR. 1997. Economics and the environment. Allen & Unwin, Sydney.
64. Woinarski, JCZ, and A Fisher. 2003. Conservation and the maintenance of biodiversity in the rangelands. Rangeland Journal 25: 157–171.
65. Young, OR. 2002. The institutional dimensions of environmental change: Fit, interplay, and scale. MIT Press, Cambridge, MA.
66. Young, MD, and B Howard. 1995. Can Australia afford a representative reserve system by 2000? Search 27: 22–26.

Chapter 6

Clarifying Competition: The Case of Wildlife and Pastoral Livestock in East Africa

Bilal Butt[1] and Matthew D Turner[2]

[1]School of Natural Resources and Environment, University of Michigan, Ann Arbor, MI 48109-1041, USA

[2]Department of Geography, University of Wisconsin, Madison, WI, 53706, USA

ABSTRACT

Contentious debates surrounding the relationship between peoples' livelihoods and protected areas in East Africa have largely revolved around claims and counter-claims about the level of competition between pastoral livestock and wildlife. Habitat and dietary overlap are often cited as the primary mechanism by which competition occurs with both overlap and lack of overlap (displacement) used as evidence of competition. Despite the importance of this issue for the economic and environmental futures of the region, there has been little scientific progress for understanding the nature of livestock–wildlife competition in pastoral landscapes. This article seeks to add conceptual clarity to this debate by focusing attention on exploitation competition in ways that are relevant to dryland East Africa. The article begins by briefly reviewing the changing understandings of the concept of competition in ecology. Requirements of competition, as defined in the literature, are then related to the ecological characteristics of East African drylands. By demonstrating that competition necessarily occurs through vegetative responses, we argue that there is the need to clarify competition by differentiating between 'proximate competition' and competition that is mediated by vegetation change across seasons. The article concludes by outlining the implications of these clarifications for the management and study of livestock–wildlife interactions.

REVIEW

Introduction

Competition between living organisms is a fundamental concept in ecology (Sommer and Worm 2002). Studies of ecological competition date as far back as the 1920s and have been used to understand species interactions and the structuring of ecological communities. These studies include those focused on theory (Connor and Simberloff 1979; Tilman 1977; Wiens 1984), mathematical modeling (Gause 1934; Lotka 1925; Mead 1967; Volterra 1926), empirical laboratory and field studies (Andrewartha and Birch 1954; Elton 1946; Harper 1977; Tilman 1981; Whittaker 1967), and application to conservation and development practices (Kreuter and Workman 1994; Vavra and Sheehy 1996).

Research that examines the relationships between wildlife and livestock has grown rapidly within the last several decades. However, there remains a great deal of controversy surrounding the characterization of wildlife–livestock relationships. Different scholars from various disciplines have suggested, implied, or stated that domestic livestock *compete* with wildlife over natural resources (Averbeck et al. 2009; Low et al. 2009; Voeten and Prins 1999; Young et al. 2005); livestock*facilitate* wildlife (Gordon 1988); livestock *co-exist or do not* *compete* with wildlife (Homewood et al. 2001; Sitters et al. 2009); or that livestock *both facilitate and compete* with wildlife (Odadi et al. 2011). Habitat and dietary overlap in the distribution of both wildlife and livestock have often been cited as the primary mechanism by which competition occurs (Beck and Peek 2005; Georgiadis et al. 2007; Madhusudan 2004; Mishra et al. 2004; Prins 1992, Prins 2000; Sitters et al. 2009; Zhongqiu et al. 2008). However, some studies have also argued that a lack of overlap (displacement) is indicative of competition (e.g., Loft et al. 1991; Stewart et al. 2002). As a result, livestock–wildlife competition can be seen as occurring with any degree of dietary or habitat overlap. While either exclusion or overlap could occur in particular situations, the persistent confusion and controversy reflect conceptual problems with how the concept is used to understand livestock–wildlife interactions. Part of the inconsistency associated with characterizing the relationships between wildlife and livestock arises from confusion over the definitions and interpretations associated with the term 'competition' (Wiens 1977, 1984).

Despite recognition of the complexity of competitive processes within the academic literature, use of the term 'competition' to characterize the relationships between wildlife and livestock is common among conservation organizations and in popular media representations of rural peoples, their livestock, and wildlife.[a] These portrayals reinforce prevailing notions that there is a zero-sum competitive relationship between livestock grazing and wildlife protection, which has contributed to policies favoring restricting livestock grazing and excluding livestock-rearing peoples near protected areas, leading to overt and covert resistance by rural people (Brockington 2002).

[a]A recent example of this appeared on the BBC's Earth News section online on June 4, 2010. Through a series of pictures depicting declining populations of Grevy's Zebra in Kenya, an accompanying storyline suggests, 'the diet of Zebra might overlap with competing livestock' (BBC 2010). Similarly, a well-known environmental news magazine cited a study in the Journal of Zoology (Ogutu et al. 2009b) whereby 'wildlife populations in Masai Mara are plummeting due to increased competition with humans and livestock' (Hance 2009).

How one conceptualizes the relationship between wildlife and livestock is enormously relevant within the East African context[b] for several reasons. First, wildlife is important to the economy of many countries, such as Kenya, where a large part of the Gross Domestic Product is derived from tourism and wildlife viewing activities (Akama 1998, Akama 2002; Waithaka 2004). Second, the megafauna that are important to the tourism industry also play an important role in ecosystem structure and functioning within the East Africa region and harbor some of the highest densities and distributions of ungulate and mammalian species in the world (Sinclair and Arcese 1995; Sinclair and Norton-Griffiths 1979; Sinclair et al. 2008). Third, livestock are important to sustain rural livelihoods in environmentally variable and heterogeneous rangelands (Ellis and Galvin 1994; Scoones 1994a). Fourth, while pastoralism is a land-use system that is potentially compatible with wildlife, there is a growing spatial overlap with a large proportion of wildlife found outside protected areas (Broten and Said 1995; Lamprey and Reid 2004; Ottichilo et al. 2000; Western et al. 2009) in areas where there are an increasing number of pastoralists (Reid et al. 2001, Reid et al. 2004). Poorly conceptualized understandings of competition have influenced the development and implementation of state and local policies associated with conservation and development, which are likely to have adverse effects on both wildlife and livestock.[c]

[b]Our primary focus in this article is on East Africa for a number of reasons. First, as we have noted above, how one conceptualizes the relationship between wildlife and livestock in east African drylands has social, political, economic, and cultural significance. Second, our backgrounds and cited references come from dryland African material. Third, East Africa is the sub-region where debates about competition among wildlife and domestic livestock are prominent and have become particularly heightened within the past decade. These tensions have shaped conservation and development initiatives in this part of the world. Finally, the East African context is drawn from our own on-going empirical investigations into wildlife–livestock competition within southern Kenya.

[c]For example, World Wildlife Fund (WWF) Biodiversity Support Program writes on their website that 'In Awash National Park, encroachment and settlement led to the illegal harvest of fuel wood and shelter wood, increased competition between wildlife and livestock, and forced many species to forage elsewhere' (Jacobs and Schloeder 2001).

The purpose of this article is therefore to *clarify the meaning of competition as it relates to wildlife and livestock in dryland Africa, particularly East Africa.*[d] This article is organized into three parts. First, we argue that unclear notions of competition circulate as common understandings without clarification due in part to its historic importance in ecology. We discuss the need to use ecological understandings that define what we mean by competition within the context of characterizing livestock–wildlife competition. Second, we move from general principles of competition to a more refined vision of wildlife-livestock competition and trace the empirical complexities of competition as mediated through vegetative response, which are shaped by climatic variability and spatial scaling. Finally, we move toward an understanding of 'exploitation competition' and outline some of the implications of conceptual clarification for measuring wildlife–livestock competition.

[d]In this article, we are primarily concerned with conservation and wildlife policies insofar as they relate to wildlife interactions with livestock in dryland environments. We are also concerned with the effects of livestock on wildlife, and not the effects of wildlife on livestock. Wildlife is defined as wild grazing ungulates and livestock as domestic cattle, sheep, goats, and camels.

Competition as an Ecological Concept

"In biological terms, the term competition is often, indeed usually, employed without definition, as if its meaning were perfectly well understood and could be taken for granted.... The result is confusion because competition means different things to different people."

*"(*Milne 1961*, pp. 40-41)"*

Over the last half-century, the application of the concept of competition has been plagued by considerable ambiguity and confusion. As Milne (1961) and Harper (1961) have both argued, the lack of conceptual clarity has been due in part to its ubiquitous nature in ecological thought. A common understanding was often assumed which did not actually exist in practice. This became particularly evident when ecological principles were applied to resource management questions (wildlife, range, fisheries management). We generally have a clear idea of competition in the abstract, but being able to identify it and measure it in the field has proven to be more complicated.

Existing research on competition between wildlife and livestock generally has treated it as*interspecific sympatric competition*, where individuals of different species vie for the same resource in an ecosystem (Prins 2000). Arguments for or against such competition have relied on dietary overlap (Fritz et al. 1996; Mishra et al. 2004; Odadi et al. 2007; Vila et al. 2009) as evidence for competition. This approach can be as simple as depicting grazers competing with grazers and browsers with browsers. More detailed analyses have relied on comparing forage species preferences (e.g., Madhusudan 2004; Odadi et al. 2011). How diets overlap may or may not be tied, in these analyses, to specific areas where wildlife and livestock graze. The geographies of wildlife and livestock movements may be ignored with conclusions of competition derived from dietary overlap and sharing common zones of movement. More sophisticated attempts have incorporated understandings of wildlife and livestock habitat to either argue for habitat overlap as evidence for ongoing competition or coexistence (Shrestha and Wegge 2008; Voeten and Prins 1999; Yoshihara et al. 2008) or no overlap as evidence for competitive exclusion, avoidance, or no possibility of competition (Loft et al. 1991; Namgail et al. 2007).

There is a need for greater clarity in applying the abstract concept of competition to livestock–wildlife interactions. References to wildlife–livestock competition elicit basic ecological and evolutionary ideas of reduced fitness, competitive exclusion, and population decline (e.g., Odadi et al. 2011; Ogutu et al. 2009b, Ogutu et al. 2011; Stewart et al. 2002). On the contrary, proponents of the 'pastoralist position' often argue for the long-term cohabitation of pastoralists (and their livestock) with wildlife (Homewood and Rodgers 1984, Homewood and Rodgers 1987; Neumann 1998). Still, meaningful competition could possibly exist without abrupt population declines. Reduced nutritional status of wildlife or domestic livestock caused by exploitation competition could lead to lower recruitment (e.g.,Hobbs et al. 1996), increased vulnerability to disease (e.g., Cleaveland et al. 2008), and emigration out of competitive areas to other areas of higher forage quality (e.g., Serneels and Lambin 2001). Making such connections in non-controlled, real-world situations, has proven very difficult since reduced fitness or population declines of livestock and wildlife are also more strongly shaped by stresses external to their competitive interactions such as droughts, disease, and habitat loss (Homewood et al. 2001). There is a strong need to delineate areas of understanding and uncertainty that surround two basic questions: in what ways can we say wildlife and livestock compete, and to what effect? Clarifying how competition is defined is an important first step toward this goal.

Competition: Definitions

In the late 1980s, Wiens provided a concise definition of ecological competition. He suggested that 'species or populations compete for resources if and when resources from a limited supply are not available or are reduced for one species due to the acts or presence of another species' (Wiens 1989, p. 3). This definition is consistent with that outlined by Milne (1961) and subsequent authors such asPrins (2000, p. 61) who outlines three conditions for interspecific competition: '(1) populations of the different species must share resources; (2) these resources must be limited, and; (3) the joint exploitation of those resources and/or interference interactions related to the resources must negatively affect the performance of either or both species'. According to Wiens (1984), competitive processes can be conceptualized in two ways: competition can occur through *interference* when resources are denied to individuals because of the aggressive actions of other individuals-or through *exploitation*, such as in the act of grazing, where limiting resources become scarcer. Although there is some evidence for wildlife avoidance of livestock and herders (e.g., Pangle and Holekamp 2010), wildlife–livestock competition is largely seen as exploitation rather than interference competition (Odadi et al. 2011).

Competition and Ecological Dynamics

Wiens (1989) highlights how competition affects individual, population, or community attributes depending on its *intensity relative to other factors shaping these attributes*. Specifically, he argued that 'competition at times may be quite intense, but nonetheless *be relatively unimportant if individual fitness or community attributes are determined largely by other factors*. Because there may be variation in the degree to which individuals in a population experience competition, some individuals may be influenced by competition without the effects of these interactions translating into population- or community-level consequences' (emphasis added) (Wiens 1989, p. 5). In fact, a trend beginning in the 1980s has been to question the equilibrium assumptions of the standard models of ecological competition, such as those of Lotka (1925), Volterra (1926), Gause (1934), and Hardin (1960), and in so doing, explain coexistence of species despite competitive interactions (Connell 1983; Schoener 1983; Sommer and Worm 2002). For example, Alley states that, 'the principle of competitive exclusion and the related assumption that communities exist at competitive equilibrium - fundamental parts of many competition theories and models - may be violated if *non-equilibrium conditions* exist in natural communities or are incorporated into competition models' (emphasis added) (Alley 1982, p. 165). In short, the effects of competition on animal populations may be small, particularly in African drylands, relative to other factors external to the competitive interaction in question.

The recognition of spatial heterogeneity has strongly shaped new views about the effect of competition on species composition. Tilman (1994) argued that a spatial model includes neighborhood competition, random dispersal among sites, and *predicts stable coexistence of a potentially unlimited number of species on a single resource*. Coexistence occurs because species with sufficiently high dispersal rates persist in sites not occupied by superior competitors. Ritchie (2002) likewise noted that it is difficult to determine whether competition occurs among mobile animals. Specifically, Richie argued that 'mobile animal species are highly likely to co-exist because of their ability to move and make choices. These choices result in resource partitioning that allow exclusive use of resources, so that the structure of communities can be predicated largely in the absence of detailed knowledge of competitive dynamics' (Ritchie 2002, p. 109). Clearly in situations such as in East Africa where livestock and wildlife are highly mobile, measuring the competition and its effects is not a straightforward problem.

Livestock–Wildlife Competition in East Africa

What is the nature of wildlife–livestock competition in East Africa? Clearly, competitive exclusion solely due to competition between mobile livestock and wildlife is unlikely. Still, a weaker form of competition undoubtedly occurs with effect. It is important to understand its nature. To do so, let's revisit the definition of ecological competition of Wiens (1989) (and adapted by Prins 2000) which requires that in order for competition to occur: (1) resources must be shared; (2) the resource must be 'limiting' to one or both of the animal populations involved; and (3) the availability of the resource must be negatively affected by one or both of the animal populations. Each of these three requirements (shared resource, limiting resource, and negative impact on resource) will be briefly discussed in the context of the competition between pastoral livestock and wildlife in East Africa.

A Shared Resource

Overlapping diets and habitat[e] are necessary but insufficient conditions for an empirical demonstration of short-term competition. Three mediating factors influence whether overlapping diets and habitat actually results in resource sharing: spatial displacement, selective grazing, and temporal displacement. Studies of grazing and browsing behavior by domestic and wild ungulates of dryland grassland regions in Africa show that grazing to be inherently patchy. Therefore, shared habitats may not result in animals grazing the same forage patches because the foraging behaviors and diets preference of the domestic and wild animals are not the same (Arnold and Dudzinsk 1978). This may be particularly true in relationships between domestic and wild ungulates as the former graze under human management while the latter may graze in areas that allow them to avoid humans (Reid et al. 2004). Even within the same patch, there is often significant but variable degrees of selectivity with respect to forage species (Manser and Brotherton 1995), phenological status of plant (Hiernaux and Turner 1996), plant part (Ayantunde et al. 1999), and physiological state of plant as affected by nutrient and moisture availability (Belsky 1986; McNaughton and Georgiadis 1986). Therefore, grazing by different ungulate species on the same patch may not effectively be sharing the same resource.[f] Moreover, grazing bouts of different species in a particular patch are most often displaced in time. Depending on the nature of the grazing, the time elapsed since grazing, and the plant community, grazing at one time of year may increase, decrease, or have little effect on the availability of a resource (a plant or plant part) during another time of year (Belsky 1986; Hiernaux and Turner 1996; McNaughton and Georgiadis 1986).

[e]Habitats refer to ecosystems with particular ecological and physical characteristics suited for a particular species to inhabit. They can be referred to in abstract or as particular places over variable spatial extent utilized by an animal population.

[f]Grazing of different forage species in the same patch may actually influence the availability of the other over time.

A Limiting Resource

Justus von Liebig's concept of a limiting resource (von Liebig 1842), while popular in the ecological and agricultural sciences, has proven difficult to implement especially in the ecosystems of dryland Africa where the biological productivity is multiply constrained with the relative limitation of a given factor varying over time and space. While animal nutrition can be limited by water, mineral availability, heat stress, infectious disease, and parasites, short-term competition between livestock and wildlife is seen as occurring through lowering the availability of fodder (Ogutu et al. 2011). Forage competition changes the energy balance of the grazing activity (units of usable energy per unit of energy expended). Grazing that reduces the availability of forage at the level of the patch may lead to greater energy expenditure by animals to maintain their intake either through less efficient selective grazing or longer movements to grazing patches[g] (Ayantunde et al. 1999). Given the relatively high abundance of quality palatable forage for both wild and domestic ungulates during the wet season, immediate competitive pressures are expected to be higher during the dry season as the energy balance of grazing animals become more limited by variations (some due to grazing) in fodder availability. Therefore, competition would be expected to be highest during the dry season when grazing ranges are constricted near available water and when overall fodder quality is lower.

[g]Given that forage species preference is strongly shaped by overall forage composition, the nutritional consequences of competition over a particular forage species may not necessarily be significantly negative since declines in the availability of one species may result in switching to another species of lower palatability but similar nutritive content.

A Negative Influence on Resource Availability

Grazing competition among domestic and wild ungulates is largely displaced in time (Prins 2000). In open range situations, it is uncommon for livestock and the wild ungulates to graze shoulder-to-shoulder on the same forage patch. Therefore, competition *necessarily* develops through the response of vegetation and soils to wildlife or livestock grazing. In the dry season, vegetation response is minimal and dry sandy soils are relatively insensitive to grazing pressure. Except in cases of heavy grazing in which the removal of thatch exposes soils to erosion from wind and heavy early rains, dry-season grazing simply reduces the stock of dry fodder available causing higher-quality forage to become less available during the same season.[h] In contrast, grazing during the rainy season has a greater chance to have longer-term interannual effects. Heavy grazing during the growing season can lead to reductions in grassland production, shifts in species composition, nutrient redistribution, and soil compaction (Augustine 2003; Hiernaux 1998; Hiernaux et al. 1999; Penning de Vries and Djitèye1982; Ruess and McNaughton 1987; Turner 1999; Veblen 2008). Therefore, while competition effects are most likely felt (e.g., a limiting resource) during the dry season, these effects are largely generated through grazing activities during the rainy season.

[h]Given the poor quality of dry grass during the dry season, dry-season grazing is selective and dry-season grazing generally does not lead to full exposure of soil (Deshmukh 1984, Deshmukh 1986).

In the dryland environments of sub-Saharan Africa, the nutrition and population dynamics of livestock (Behnke et al. 1993; Ellis and Swift 1988; Illius and O'Connor 1999; Sullivan and Rohde 2002) and wildlife (Hopcraft et al. 2010; Ogutu et al. 2009a) are strongly affected by factors external to the grazer–forage relationship, most particularly climate. Whether conceptualized as non-equilibrium systems or not, wildlife–livestock competition, as mediated through vegetative response to grazing, may have a small effect relative to the high variability in conditions affecting vegetation, livestock, and wildlife. There are many factors influencing domestic and wild animal populations external to the competitive relations among grazing animals. For example, studies of domestic livestock (e.g., Ellis and Swift 1988; Scoones 1993, Scoones 1994b; Sullivan and Rohde

2002) and wildlife (Georgiadis et al. 2003, Georgiadis et al. 2007) population dynamics in dryland Africa find that they are often 'density independent' with variations in population growth rate unrelated to the size of the population. Moreover, nutritional studies also show that the seasonal variation in the quality of the vegetation often is a more important factor affecting animal nutrition than the stocking density of the broad areas over which wild and domestic ungulates circulate across the year (e.g.,Ayantunde et al. 1999; Fernández-Rivera et al. 2005; Milner-Gulland et al. 2011). Dryland vegetative productivity has been found to be shaped less by recent grazing history (unless very heavy) than by longer-term grazing history coupled with seasonal and interannual rainfall variation (e.g., Hiernaux and Turner 1996; Penning deVries and Djitèye 1982; Turner 1999). In sum, if one ignores the strong influence of rainfall variation in dryland systems, one is likely to over-estimate the relative importance of competition in affecting livestock or wildlife. Many conservation and development policies often fail to recognize the seasonal variability in rainfall, which strongly affects forage growth (Butt 2010a). This variability also strongly influences the resource access strategies of domestic livestock (Butt 2010b). There is little acknowledgment of seasonal variation and differential space–time utilization of forage resources by both wildlife and livestock in the few published conservation planning initiatives.

Unpacking Exploitation Competition in East Africa

In order to effectively move beyond conceptualizations of competition as related to overlap and/or displacement, there is the need to further unpack the notion of exploitation competition in a way that is relevant to dryland East Africa. A necessary condition for 'exploitation competition' is that the actions of one population meaningfully limit the availability of forage for the other population. The high seasonality of vegetation productivity in rangeland environments of East Africa (Butt 2010a;Deshmukh 1986) strongly shapes the potential for competition. It is rare for grazing-induced forage shortage to significantly limit ungulate nutrition during the wet season when vegetation is most sensitive to grazing (Voeten and Prins 1999). Vegetation is much more resilient to dry season grazing when forage is limiting to grazing ungulates (McNaughton 1985, Mworia et al. 1997)Therefore, we wish to distinguish two types of exploitation competition, which can occur during the dry season.

The first is the grazing competition for quality forage as the dry season progresses and the availability of digestible forage declines. Competition occurs through the direct effect of one population consuming what the other population would have consumed within later weeks. This is most close to current conceptualizations of competition held by wildlife managers - one ungulate's consumption literally takes quality forage away from another.[i] This more proximate form of exploitation competition is necessarily associated with grazing overlap between wildlife and domestic livestock in space and time. This overlap can occur at the scale of weeks to months. It is most common where grazing movements are restricted to those areas near more permanent water sources and pastoral settlements. Drought conditions can exacerbate these competitive pressures through the reduction in surface water and/or forage. Proximate competition occurs during periods and at places where quality forage availability is already limited, whether or not a competing population is present. Under these situations, all ungulate species are generally competing between and among each other, and the amount of forage removed through the actions of others is very small relative to annual forage requirements.[j] Nonetheless, a decline in a population's access to nutritional quality may have, in difficult years, a significant effect on survivorship.

[i]Termites may as well be important competitors with both livestock and wild ungulates for dry vegetation (e.g., Hopcraft 2000, p. 271; Newsome 1971).

[j]At sites where the quality of dry forage is ubiquitously low, grazing by a population may not have a significantly negative effect on the nutrition of another since it is the quality of the forage that is most limiting not its quantity. Therefore, we may see the potential for proximate dry-season competition to decline as the dry season progresses and the heterogeneity of forage quality declines.

The second type of exploitation competition is more displaced in time. Given the important effects of rainy-season grazing on the quality and quantity of forage production (as influenced by species composition and productivity of rangelands), a major mechanism through which competition could occur in drylands is the reduction of forage availability during the dry season caused by wet-season grazing. These conditions can exist within the same year or interannually. In these cases, competition is displaced over time. The negative impacts of this form of competition on forage availability are potentially much larger than the effects of proximate competition but they have proven to be more difficult to empirically document.

Implications for Future Research

Given the porosity of their borders, wildlife and livestock graze inside and outside of protected areas in East Africa (Peden 1987; Reid et al. 2001, Reid et al. 2003; Western 1982; Western et al. 2009). We therefore need to build our understanding of wildlife–livestock competition to better manage the mixed grazing, farming, and wildlife landscapes of East Africa. Despite common misperceptions, evidence, as described above, suggests that wildlife–livestock competition does not lead to competitive exclusion and may have a smaller impact on wildlife and livestock populations than factors external to the wildlife–livestock interaction. Other factors such as drought, habitat/pasture loss due to agriculture, disease, and hunting/predation are likely to more significantly influence the viability of wildlife and livestock populations (Homewood et al. 2001). Still competition undoubtedly occurs with an effect on the grazing strategies and grazing efficiencies of these populations. As discussed previously, we cannot infer competition simply from the sharing of a preferred resource or space (dietary or habitat overlap). The resource must be the same, limiting, and negatively affected by the activities of the other species. Where attempts have been made to move beyond diet/habitat overlap, researchers have relied on experimental designs where wildlife and livestock are enclosed in fenced pastures. For example, the effects of wildlife and livestock on vegetation are often conducted through timed runs within experimental blocks measuring a few hectares (Odadi et al. 2007; Young et al. 1998). While such work provides insights on the mechanisms behind competition, they provide little information about the nature of wildlife–livestock interactions in less-controlled, real-world situations where pastoral livestock and wildlife are highly mobile and the variability of external factors are strongly felt.

Given these circumstances, there is a strong need to combine the insights of such experimental work with field studies that more seriously address the complex contexts within which wildlife–livestock competition occurs. Our review points to a number of features that new research should incorporate.

First, empirical analysis of wildlife–livestock competition requires more thorough analyses of how 'proximate competition' influences wildlife and domestic livestock nutrition as forage scarcity increases during the dry season. There are significant questions with respect to interspecific grazing interactions, selective grazing, and changing heterogeneity of forage quality in pasture areas around the key resources relied upon by domestic livestock and wildlife populations during the dry season (Ngugi and Conant 2008). Animal nutrition and range ecology work focused on this question should be performed around key seasonally variable resources used by wildlife and livestock.

Second, wildlife and livestock competition is most likely to be displaced in time as mediated through vegetative response to rainy-season grazing. Wildlife–livestock competition research, which has been dominated by wildlife ecologists, must more seriously address range ecology. In particular, future research must address how both forage quality and quantity change with respect to differential grazing pressures within and across seasons. Given that rangelands are characterized by spatially and temporally heterogeneous landscapes with a variety of different land covers, soil types, and rainfall gradients, there is the need to understand how these factors influence above ground biomass and forage quality.

Third, there is a need for greater spatial specificity of livestock and wildlife grazing and browsing patterns to understand their true overlap across time. While radio collaring has been a standard approach in wildlife ecology for some time, more spatially refined documentation of the grazing 'orbits' of pastoral-managed livestock has only recently been developed (Butt 2010b, 2011;Coppolillo 2000). Moreover, pastoralists or community groups themselves will provide very useful information about landscape change and the interactions of domestic and wild ungulates in the bush (Goldman 2007; Low et al. 2009; Msoffe et al. 2010; Oba and Kaitira 2006).

Finally, wildlife–livestock interaction needs to be studied in particular places given the temporal displacement inherent to nutritional consequences of grazing. Grazing occurs in different places at different times of the year. Vegetative response is most sensitive to grazing during the rainy season, while forage is most limited during the dry season. Therefore, future research should aim to monitor grazing and vegetative response across at least an annual cycle, rather than solely during dry periods.

CONCLUSIONS

In this article, we have discussed the importance that wildlife and livestock play in the ecology and economy of East African countries. A review of the scientific literature suggests that the term 'competition' has been used to characterize the relationships between wildlife and pastoral livestock and reveals that competition has been perceived as occurring with any degree of habitat or dietary overlap. We have demonstrated how characterizing the relationships between wildlife and pastoral livestock as competition is problematic given the vague definitions and contexts associated with the concept. To help clarify 'competition', we have traced the history, definitions, and ecological dynamics associated with the concept. We then discuss the three conditions necessary for competition to occur (a shared resource, a limiting resource, and a negative influence on resource availability) within the context of pastoral managed livestock and wildlife in dryland regions of Africa. This literature suggests that wildlife–livestock competition is largely exploitative. However, exploitation competition is mediated through vegetative responses to grazing, which are shaped by climatic variability, grazing selectivity, and patchiness, which vary across heterogeneous landscapes in both time and space.

Both wild ungulates and domestic livestock are mobile and competition between them can only occur if their grazing occurs at the same place. Vegetative change is most associated with grazing during the growing season - a time when forage availability is less limiting to animal nutrition. Therefore, competition that is seasonally displaced (grazing in one season affects vegetation and animal nutrition in subsequent seasons) is the most prevalent form of competition but at the same time the most subtle and difficult to investigate. Evidence of such competition necessarily is vegetative change - requiring greater understanding of vegetative ecology than is typical for wildlife managers. We have argued that in order to move beyond conceptualizations of competition that are rooted in dietary overlap and/or displacement, there is the need to clarify exploitation competition by differentiating between competition that occurs over a fixed resource or 'proximate competition' and that which is mediated through vegetative response to grazing at an earlier time. The former is associated with forage reductions during the dry season when there is not vegetative growth, while the latter is most associated with wet season grazing where the competitive effects of grazing are mediated by vegetative response (changes in species composition and productivity) and is felt across seasons. We end the article by providing a number of implications for future research, which should be focused on less controlled, real-world situations. These recommendations include focusing on the nutritional implications for wildlife and livestock as forage quantity decreases; the vegetative responses to grazing within and across seasons; greater spatial specificity of wildlife and livestock grazing patterns; and specific grazing locations across the entire seasonal calendar.

Increased rigor in thinking about the competition among wildlife and livestock not only leads to the identification of important areas for further research (areas of uncertainty), but also reveals areas of relative certainty to guide policy particularly related to wildlife protection and management. Contrary to popular impression, the presence or absence of dietary or habitat overlap is insufficient to claim competition between livestock or wildlife. Such arguments provide little room for mutually beneficial use of common rangeland resources by wildlife and livestock for which there is historical and contemporary evidence (Homewood and Rodgers 1987; Little 1996). Such thinking not only leads to over-estimations of competition, but also leads to land-use decisions that exclude grazing wildlife from pasturelands or livestock from protected areas - decisions that will increase the vulnerability of the whole wildlife–livestock–grassland system.

AUTHORS' CONTRIBUTIONS

With BB taking the lead particularly with respect to the literature review, both authors (BB and MT) contributed significantly to the conceptual framing and argumentation of this article. Both authors read and approved the final manuscript.

ACKNOWLEDGMENTS

We would like to thank Augustine Ayantunde for helpful comments on earlier parts of this manuscript.

REFERENCES

1. Akama, JS. 1998. The evolution of wildlife conservation policies in Kenya. Journal of Third World Studies 15: 103–117.
2. Akama, JS. 2002. The role of Government in the development of tourism in Kenya. International Journal of Tourism Research 4: 1–14.
3. Alley, TR. 1982. Competition theory, evolution, and the concept of an ecological niche. Acta Biotheor 31: 165–179.
4. Andrewartha, HG, and C Birch. 1954. The distribution and abundance of animals. University of Chicago Press, Chicago.
5. Arnold, GW, and ML Dudzinsk. 1978. Ethology of free-ranging domestic animals. Developmens in Animal and Veterinary Science 2. Elsevier, Amsterdam.
6. Augustine, DJ. 2003. Long-term, livestock-mediated redistribution of nitrogen and phosphorus in an East African savanna. J Appl Ecol 40: 137–149.
7. Averbeck, C, A Apio, M Plath, and T Wronski. 2009. Environmental parameters and anthropogenic effects predicting the spatial distribution of wild ungulates in the Akagera savannah ecosystem. Afr J Ecol 47: 756–766.
8. Ayantunde, A, P Hiernaux, S Fernandez-Rivera, H van Keulen, and HMJ Udo. 1999. Selective grazing by cattle on spatially and seasonally heterogeneous rangeland in Sahel. J Arid Environ 42: 261–279.
9. BBC. 2010. In Pictures: Grevy's zebra. http://news.bbc.co.uk/earth/hi/earth_news/newsid_8719000/8719473.stm *webcite* Accessed 08 Nov 2010
10. Beck, JL, and JM Peek. 2005. Diet composition, forage selection, and potential for forage competition among elk, deer, and

livestock on aspen-sagebrush summer range. Rangel Ecol Manage 58: 135–147.

11. Behnke, RH, I Scoones, and C Kerven. 1993. Range ecology at disequilibrium : New models of natural variability and pastoral adaptation in African savannas. Overseas Development Institute, London.
12. Belsky, AJ. 1986. Does herbivory benefit plants? A review of the evidence. Am Nat 127: 870–892.
13. Brockington, D. 2002. Fortress conservation: The preservation of the Mkomazi Game Reserve, Tanzania. James Currey, Oxford.
14. Broten, M, and M Said. 1995. Population trends of ungulates in and around Kenya's Masai Mara Reserve. In Serengeti II: Dynamics, management and conservation of an ecosystem, ed. Sinclair ARE, Arcese P, 169–193. University of Chicago Press, Chicago.
15. Butt, B. 2010. Pastoral resource access and utilization: Quantifying the spatial and temporal relationships between livestock mobility, density and biomass availability in southern Kenya. Land Degradation and Development 21: 520–539.
16. Butt, B. 2010. Seasonal space-time dynamics of cattle behavior and mobility among Maasai pastoralists in semi-arid Kenya. J Arid Environ 74: 403–413.
17. Butt, B. 2011. Coping with uncertainty and variability: The influence of protected areas on pastoral herding strategies in East Africa. Hum Ecol 39: 289–307.
18. Cleaveland, S, C Packer, K Hampson, M Kaare, R Kock, M Craft, T Lembo, T Mlengeya, and A Dobson. 2008. The multiple roles of infectious diseases in the serengeti ecosystem. In In Serengeti III: Human impacts on dynamic ecosystems, ed. Sinclair ARE, Packer C, Mduma S, Fryxell JM, 209–240. University of Chicago Press, Chicago.
19. Connell, JH. 1983. On the prevalence and relative importance of interspecific competition: Evidence from field experiments. American Naturality 122: 661–696.
20. Connor, EF, and D Simberloff. 1979. The assembly of species communities; Chance or competition. Ecology 60: 1132–1140.

21. Coppolillo, PB. 2000. The landscape ecology of pastoral herding: Spatial analysis of land use and livestock production in East Africa. Hum Ecol 28: 527–560.
22. Deshmukh, I. 1984. Common relationship between precipitation and grassland peak biomass for East and Southern Africa. Afr J Ecol 22: 181–186.
23. Deshmukh, I. 1986. Primary production of a grassland in Nairobi National Park, Kenya. J Appl Ecol 23: 115–123.
24. Ellis, J, and KA Galvin. 1994. Climate patterns and land-use practices in the dry zones of Africa. Bioscience 44: 340–349.
25. Ellis, J, and DM Swift. 1988. Stability of African pastoral ecosystems: Alternate paradigms and implications for development. J Range Manage 41: 450–459.
26. Elton, CS. 1946. Competition and the structure of ecological communities. J Anim Ecol 15: 54–68.
27. Fernández-Rivera, S, P Hiernaux, TO Williams, MD Turner, E Schlecht, and A Salla. 2005. Nutritional constraints to grazing ruminants in the millet-cowpea-livestock farming system of the Sahel. In Coping with feed scarcity in smallholder livestock systems in developing countries, ed. Ayantunde AA, Fernández-Rivera S, McCrabb G, 157–182. International Livestock Research Institute, Nairobi.
28. Fritz, H, M de Garine-Wichatitsky, and G Letessier. 1996. Habitat use by sympatric wild and domestic herbivores in an African savanna woodland: The influence of cattle spatial behaviour. J Appl Ecol 33: 589–598.
29. Gause, GF. 1934. The struggle for existence. Williams ?& Wilkins Company, Baltimore.
30. Georgiadis, N, M Hack, and K Turpin. 2003. The influence of rainfall on zebra population dynamics: implications for management. J Appl Ecol 40: 125–136.
31. Georgiadis, N, JGN Olwero, G Ojwang, and SS Romañach. 2007. Savanna herbivore dynamics in a livestock-dominated landscape: I. Dependence on land use, rainfall, density, and time. Biol Conserv 137: 461–472.
32. Goldman, M. 2007. Tracking wildebeest, locating knowledge: Maasai and conservation biology understandings of wildebeest

behavior in Northern Tanzania. Environment and Planning D Society and Space 25: 307–331.

33. Gordon, IJ. 1988. Facilitation of red deer grazing by cattle and its impact on red deer performance. J Appl Ecol 25: 1–9.
34. Hance, J. 2009. Famous Kenyan park experiencing large declines in wildlife. http://news.mongabay.com/2009/0421-hance_masaimara.html *webcite* Accessed 21 Aug 2011
35. Hardin, G. 1960. The competitive exclusion principle. Science 131: 1292–1297. PubMed Abstract |
36. Harper, D. 1961. Approaches to the study of plant competition. In Mechanisms in biological competition symposia of the society for experimental biology, XV, 1-39. Cambridge University Press, Cambridge.
37. Harper, D. 1977. The population biology of plants. Academic Press, New York.
38. Hiernaux, P. 1998. Effects of grazing on plant species composition and spatial distribution in rangelands of the Sahel. Plant Ecology 138: 191–202.
39. Hiernaux, P, CL Bielders, C Valentin, A Bationo, and S Fernandez-Rivera. 1999. Effects of livestock grazing on physical and chemical properties of sandy soils in Sahelian rangelands. J Arid Environ 41: 231–245.
40. Hiernaux, P, and MD Turner. 1996. The effect of clipping on growth and nutrient uptake of Sahelian annual rangelands. J Appl Ecol 33: 387–399.
41. Hobbs, NT, DL Baker, GD Bear, and DC Bowden. 1996. Ungulate grazing in Sagebrush grassland: Effects of resource competition on secondary production. Ecol Appl 6: 218–227.
42. Homewood, K, EF Lambin, E Coast, A Kariuki, I Kikula, J Kivelia, M Said, S Serneels, and M Thompson. 2001. Long-term changes in Serengeti-Mara wildebeest and land cover: Pastoralism, population, or policies? Proc Natl Acad Sci U S A 98: 12544–12549.
43. Homewood, K, and W Rodgers. 1987. Pastoralism, conservation and the overgrazing controversy. In Conservation in Africa, ed. Anderson D, Grove R, 111–128. Cambridge University Press, Cambridge.

44. Homewood, K, and W Rodgers. 1984. Pastoralism and conservation. Hum Ecol 12: 431–441.
45. Hopcraft, D. 2000. Wildlife land use and the great experiment. In Wildlife conservation by sustainable use, ed. Prins HHT, Grootenhuis JG, Dolan TT, 267–275. Kluwer Academic Publishers, Boston.
46. Hopcraft, JGC, H Olff, and ARE Sinclair. 2010. Herbivores, resources and risks: Alternating regulation along primary environmental gradients in savannas. Trends Ecol Evol 25: 119–128.
47. Illius, AW, and TG O'Connor. 1999. On the relevance of nonequilibrium concepts to arid and semiarid grazing systems. Ecol Appl 9: 798–813.
48. Jacobs, MJ, and CA Schloeder. 2001. Impacts of conflict on biodiversity and protected areas in Ethiopia. Biodiversity Support Program.WWF, Washington, DC.
49. Kreuter, UP, and JP Workman. 1994. Government policy effects on cattle and wildlife ranching profits in Zimbabwe. J Range Manage 47: 264–269.
50. Lamprey, RH, and RS Reid. 2004. Expansion of human settlement in Kenya's Maasai Mara: What future for pastoralism and wildlife? J Biogeogr 31: 997–1032.
51. Little, PD. 1996. Pastoralism, biodiversity, and the shaping o f savanna landscapes in East Africa. Africa 66: 37–51.
52. Loft, ER, JW Menke, and JG Kie. 1991. Habitat shifts by mule deer: The influence of cattle grazing. Journal of Wildlife Management 55: 16–26.
53. Lotka, AJ. 1925. Elements of physical biology. Williams ?& Wilkins Company, Baltimore.
54. Low, B, SR Sundaresan, IR Fischhoff, and DI Rubenstein. 2009. Partnering with local communities to identify conservation priorities for endangered Grevy's zebra. Biol Conserv 142: 1548–1555.
55. Madhusudan, MD. 2004. Recovery of wild large herbivores following livestock decline in a tropical Indian wildlife reserve. J Appl Ecol 41: 858–869.

56. Manser, MB, and PNM Brotherton. 1995. Environmental constraints on the foraging behavior of a Dwarf Antelope (Madoqua-Kirkii). Oecologia 102: 404–412.
57. McNaughton, SJ. 1985. Ecology of a Grazing System: The Serengeti. JMonographs 55: 259–294.
58. McNaughton, SJ, and NJ Georgiadis. 1986. Ecology of African grazing and browsing mammals. Annual Review of Ecology and Systematics 17: 39–65.
59. Mead, R. 1967. A mathematical model for the estimation of inter-plant competition. Biometrics 23: 189–205.
60. Milne, A. 1961. Definition of competition among animals. In Mechanisms in biological competition symposia of the society for experimental biology XV, 40-61. Cambridge University Press, Cambridge.
61. Milner-Gulland, EJ, JM Fryxell, and ARE Sinclair. 2011. Animal migration: A synthesis. Oxford University Press, Oxford.
62. Mishra, C, HHT Prins, SE Van Wieren, P Ketner, and IMA Heitkönig. 2004. Competition between domestic livestock and wild bharal Pseudois nayaur in the Indian Trans-Himalaya. J Appl Ecol 41: 344–354.
63. Msoffe, FU, JO Ogutu, J Kaaya, C Bedelian, MY Said, SC Kifugo, RS Reid, M Neselle, P van Gardingen, and S Thirgood. 2010. Participatory wildlife surveys in communal lands: a case study from Simanjiro, Tanzania. Afr J Ecol 48: 727–735.
64. Mworia, JK, WM Mnene, DK Musembi, and RS Reid. 1997. Reselienceof soils and vegetation subjected to different grazing intensities in a semi-arid Tangeland of Kenya. African Journal of Range ?& Forage Science 14: 26–31.
65. Namgail, T, JL Fox, and YV Bhatnagar. 2007. Habitat shift and time budget of the Tibetan argali: The influence of livestock grazing. Ecological Research 22: 25–31.
66. Neumann, R. 1998. Imposing wilderness: Struggles over livelihood and nature preservation in Africa. University of California Press, Berkeley.
67. Newsome, AE. 1971. Competition between wildlife and domestick livestock. Aust Vet J 47: 577–586.

68. Ngugi, MK, and RT Conant. 2008. Ecological and social characterization of key resource areas in Kenyan rangelands. J Arid Environ 72: 820–835.
69. Oba, G, and LM Kaitira. 2006. Herder knowledge of landscape assessments in arid rangelands in northern Tanzania. J Arid Environ 66: 168–186.
70. Odadi, WO, M Karachi, S Abdulrazak, and TP Young. 2011. African wild ngulates compete with or facilitate cattle depending on season. Science 333: 1753–1755.
71. Odadi, WO, TP Young, and JB Okeyo-Owuor. 2007. Effects of wildlife on cattle diets in Laikipia rangeland, Kenya. Rangel Ecol Manage 60: 179–185.
72. Ogutu, J, HP Piepho, HT Dublin, N Bhola, and R Reid. 2009. Rainfall extremes explain interannual shifts in timing and synchrony of calving in topi and warthog. Population Ecology 52: 89–102.
73. Ogutu, J, HP Piepho, HT Dublin, N Bhola, and RS Reid. 2009b. Dynamics of Mara Serengeti ungulates in relation to land use changes. J Zool 278: 1–14.
74. Ogutu, JO, N Owen-Smith, HP Piepho, and MY Said. 2011. Continuing wildlife population declines and range contraction in the Mara region of Kenya during 1977–2009. J Zool 285: 99–109.
75. Ottichilo, WK, MY Said, J de Leeuw, AK Skidmore, and HHT Prins. 2000. Population trends of large non-migratory wild herbivores and livestock in the Masai Mara ecosystem, Kenya, between 1977 and 1997. Afr J Ecol 38: 202–216.
76. Pangle, WM, and KE Holekamp. 2010. Lethal and nonlethal anthropogenic effects on spotted hyenas in the Masai Mara National Reserve. J Mammal 91: 154–164.
77. Peden, DG. 1987. Livestock and wildlife population distributions in relation to aridity and human populations in Kenya. J Range Manage 40: 67–71.
78. Penning de Vries, FWT, and MA Djitèye. 1982. La productivité des pâturages sahéliens. Centre for Agricultural Publishing and Documentation, Wageningen. PubMed Abstract

79. Prins, HHT. 1992. The pastoral road to extinction: Competition between wildlife and traditional pastoralism in East Africa. Environmental Conservation 19: 117–123.

80. Prins, HHT. 2000. Competition between wildlife and livestock in Africa. In Wildlife conservation by sustainable use, ed. Prins HHT, Grootenhuis JG, Dolan TT, 51–80. Kluwer Academic Publishers, Boston.

81. Reid, R, M Rainy, J Ogutu, R Kruska, M McCartney, M Nyabenge, K Kimani, M Kshatriya, J Worden, L Ng'ang'a, J Owuor, J Kinoti, E Njuguna, C Wilson, and R Lamprey. 2003. People, wildlife and livestock in the Mara ecosystem: the Mara count 2002, Report, Mara Count 2002. International Livestock Research Institute, Nairobi.

82. Reid, R, PK Thornton, and R Kruska. 2004. Loss and fragmentation of habitat for pastoral people and wildlife in east Africa: Concepts and issues. African Journal of Range and Forage Science 21: 171–181.

83. Reid, R, ME Rainy, CJ Wilson, E Harris, RL Kruska, MN Waweru, SA MacMillan, and JS Worden. 2001. Wildlife cluster around pastoral settlements in Africa, creating multiple species associations of wildlife, livestock and people, PLE Science Series # 2. International Livestock Research Institute, Nairobi.

84. Ritchie, ME. 2002. Competition and coexistence of mobile animals. In Competition and coexistence, ed. Sommer U, Worm B, 109–132. Springer, Berlin.

85. Ruess, RW, and SJ McNaughton. 1987. Grazing and the dynamics of nutrient and energy regulated microbial processes in the Serengeti Grasslands. Oikos 49: 101–110.

86. Schoener, TW. 1983. Field experiments on interspecific competition. Am Nat 122: 240–285.

87. Scoones, I. 1993. Why are there so many animals? Cattle population dynamics in the communal areas of Zimbabwe. In Range ecology at disequilibrium: new models of natural variability and pastoral adaptation in African savannas, ed. Behnke R, Kerven C, 62–76. Overseas Development Institute, London.

88. Scoones, I. 1994. Living with uncertainty: new directions in pastoral development in Africa. Intermediate Technology Publications, London.
89. Scoones, I. 1994. New directions in pastoral development in Africa. In Living with uncertainty: New directions in pastoral development in Africa, ed. Scoones I, 1–36. Intermediate Technology Publications Ltd., London.
90. Serneels, S, and EF Lambin. 2001. Impact of land-use changes on the wildebeest migration in the northern part of the Serengeti-Mara ecosystem. J Biogeogr 28: 391–407.
91. Shrestha, R, and P Wegge. 2008. Wild sheep and livestock in Nepal Trans-Himalaya: Coexistence or competition? Environmental Conservation 35: 125–136.
92. 2008. Serengeti III: Human impacts on ecosystem dynamics,Sinclair ARE, Packer C, Mduma S, Fryxell JM University of Chicago Press, Chicago.
93. 1995. Serengeti II: Dynamics, management, and conservation of an ecosystem,Sinclair ARE, Arcese P University of Chicago Press, Chicago.
94. 1979. Serengeti, dynamics of an ecosystem,Sinclair ARE, Norton-Griffiths M University of Chicago Press, Chicago.
95. Sitters, J, IMA Heitnig, M Holmgren, and GSO Ojwang. 2009. Herded cattle and wild grazers partition water but share forage resources during dry years in East African savannas. Biol Conserv 142: 738–750.
96. Sommer, U, and B Worm. 2002. Competition and coexistence. Springer-Verlag, Berlin.
97. Stewart, KM, RT Bowyer, JG Kie, NJ Cimon, BK Johnson, and TJ O'Shea. 2002. Temporospatial distributions of elk, mule deer, and cattle: Resource partitioning and competitive displacement. J Mammal 83: 229–244.
98. Sullivan, S, and R Rohde. 2002. On non-equilibrium in arid and semi-arid grazing systems. J Biogeogr 29: 1595–1618.
99. Tilman, D. 1977. Resource competition between planktonic algae: Experimental and theoretical approach. Ecology 58: 338–348.

100. Tilman, D. 1981. Tests of resource competition theory using four species of Lake Michigan Algae. Ecology 62: 802–815.
101. Tilman, D. 1994. Competition and biodiversity in spatially structured habitats. Ecology 75: 2–16.
102. Turner, MD. 1999. Spatial and temporal scaling of grazing impact on the species composition and productivity of Sahelian annual grasslands. J Arid Environ 41: 277–297.
103. Vavra, M, and D Sheehy. 1996. Improving elk habitat characteristics with livestock grazing. Rangelands 18: 182–185.
104. Veblen, KE. 2008. Season- and herbivore-dependent competition and facilitation in a semiarid savanna. Ecology 89: 1532–1540.
105. Vila, AR, L Borrelli, and L Martinez. 2009. Dietary overlap between huemul and livestock in Los Alerces National Park, Argentina. Journal of Wildlife Management 73: 368–373.
106. Voeten, MM, and HHT Prins. 1999. Resource partitioning between sympatric wild and domestic herbivores in the Tarangire region of Tanzania. Oecologia 120: 287–294.
107. Volterra, V. 1926. Variazioni e fluttuazioni del numero d'individui in specie animali conviventi. Memoria della Regia Accademia Nazionale dei Lincei, ser. 6 (2): 31–113.
108. von Liebig, J. 1842. Chemistry in its application to agriculture and physiology. Taylor ?& Walton, London.
109. Waithaka, J. 2004. Maasai Mara: An ecosystem under siege: an African case study on the societal dimension of rangeland conservation. African Journal of Range and Forage Science 21: 79–88.
110. Western, D. 1982. Amboseli National Park: Enlisting landowners to conserve migratory wildlife. Ambio 11: 302–308.
111. Western, D, S Russell, and I Cuthill. 2009. The status of wildlife in protected areas compared to non-protected areas of Kenya. PLoS One 4: 1–6.
112. Whittaker, RH. 1967. Gradient analysis of vegetation. Biol Rev 42: 207–264. PubMed Abstract |
113. Wiens, JA. 1977. Competition and variable environments. Am Sci 65: 590–597.

114. Wiens, JA. 1984. On understanding a nonequilibrium world: Myth and reality in community patterns and processes. In Ecological communities: Conceptual issues and the evidence, ed. Strong DR, Simberloff D, Abele LG, Thistle AB, 439–457. Princeton University Press, Princeton.

115. Wiens, JA. 1989. The ecology of bird communities. Cambridge University Press, Cambridge.

116. Yoshihara, Y, TY Ito, B Lhagvasuren, and s Takatsuki. 2008. A comparison of food resources used by Mongolian gazelles and sympatric livestock in three areas in Mongolia. J Arid Environ 72: 48–55.

117. Young, TP, B Okello, D Kinyua, and T Palmer. 1998. KLEE: A long-term multi-species herbivore exclusion experiment in Laikipia, Kenya. African Journal of Range and Forage Science 14: 92–104.

118. Young, TP, T Palmer, and ME Gadd. 2005. Competition and compensation among cattle, zebras, and elephants in a semi-arid savanna in Laikipia, Kenya. Biol Conserv 122: 351–359.

119. Zhongqiu, LI, J Zhigang, and LI Chunwang. 2008. Dietary overlap of Przewalski's Gazelle, Tibetan Gazelle, and Tibetan Sheep on the Qinghai-Tibet Plateau. Journal of Wildlife Management 72: 944–948.

Chapter 7

Envisioning the Sustainability of the Production of Short Rotation Coppice on Grassland

Christine Rösch[1], Cisco Aust[2], and Juliane Jörissen[1]

[1]Institute for Technology Assessment and Systems Analysis (ITAS), Karlsruhe Institute of Technology (KIT), Karlsruhe, Germany

[2]Federal Forest Research Institute Baden-Wuerttemberg (FVA), Freiburg, Germany

ABSTRACT

Background

Short rotation coppices (SRC) offer a chance to enhance sustainable biomass production and to enlarge domestic wood resources. On arable land, SRC are less competitive in many regions compared to cropping systems and thus farmers are interested to establish SRC on grassland. This change of land use is discussed controversially because of the associated effects on the carbon-sequestering ecosystem grassland. In this paper, land use changes will be analyzed in reference to sustainability objectives, including aspects of land use competition with food and energy production or the preservation of biodiversity.

Methods

The Integrative Sustainability Concept has been adapted and applied to identify the relevant sustainability objectives using a holistic approach in order to assess the potential for the change in sustainable land use towards an expanded supply of lignocellulose-based bioenergy.

Results

A set of sustainability criteria with a special focus on the change in land use from grassland to SRC has been developed and discussed. For selected criteria, adequate sustainability indicators have been chosen and applied to a case study of the state of Baden-Wuerttemberg in Germany. The indicators have been implemented by performing a spatial differentiated bottom-up assessment using GIS-based data. The results revealed that around 13% of the grassland could be converted into SRC without harming the considered sustainability criteria.

Conclusions

The exploitation of the assessed potential for the change in sustainable land use from grassland to SRC could contribute to decentralize renewable energy supply, local employment and income. With regard to the economic values otherwise lost-, it is recommended not to set grassland conservation as an area-wide imperative. Instead strategies to direct the establishment of SRC to the most suitable areas and technologies should be developed realizing that land use transformation will be performed in a minimal invasive way to ensure compliance with the environment and climate protection.

BACKGROUND

The European Union (EU) has acknowledged the ecological functions of grassland and introduced measures within the Common Agricultural Policy (CAP) for its protection against conversion into arable land as well as for preserving or enhancing its ecological quality. Cross compliance, linking the full receipt of direct payments for farmers to compliance with environmental and further standards, explicitly addresses the protection of grassland (Regulation (EC) nos. 796/2004 and 73/2009). EU member states must ensure that at the national scale, the ratio of the land under permanent grassland in relation to the total agricultural area may not decrease by more than 10% compared to the year 2003. Germany implemented precautionary measures before the 10% threshold has been reached. A need for authorization to convert grassland to arable land is required if the 5% level is exceeded.

Resulting from the introduction of cross compliance in Germany, the ratio of permanent grassland is declining. By 2010, the 5% threshold was exceeded in four federal states. The federal states level is the level on which this ratio is calculated in Germany [1]. As a result, for any further conversion of permanent grassland, authorization has to be sought and as a rule, grassland has to be created elsewhere as compensation. More incentives for grassland are provided at the regional level in order to preserve certain management practices of extensive grassland. For less favoured areas, natural handicap payments are given to support grassland management even under adverse conditions. However, these incentives could not stop the further decline of grassland either.

The main driving forces for the decrease of grassland are not only the continuously shrinking cattle stock being the primary user of grassland [2], but also the German Renewable Energies Act which makes the production of maize as a feedstock for biogas plants profitable. Around 53% of the permanent grassland that has been converted to arable land is being used to cultivate maize [1]. This land use change can induce negative impacts on the environment because grassland provides several important ecological benefits including carbon storage, protection of soil from erosion, ground water formation, and habitat function. Besides, grassland is forming an essential part of the cultural landscapes serving as a basis for nature-related recreation and tourism.

Since July 2011, the federal state of Baden-Wuerttemberg has prohibited the conversion of grassland by law. In the EU, the absolute maintenance of grassland is under discussion and could become one of the pillars of the projected ecologization of the CAP. Such a regulation would prevent any type of changes in grassland use regardless of their impacts. The justifications for such an absolute prohibition of grassland conversion are the multifunctional environmental benefits of grassland. Thereby the focus is on the functions of grassland as a host of endangered wildlife and plants and as a protector of huge carbon deposits which would otherwise be released as greenhouse gases. In Baden-Wuerttemberg, around 6% of the permanent grassland covering 532,000 ha or 38% of the agricultural used land is extensively managed, e.g. as meadow with scattered fruit trees [3]. This area is of great importance for the preservation of biodiversity. Part of the remaining grassland could be used for alternative biomass production. Permanent crops such as *Miscanthus* or fast growing trees have higher biomass yields compared to grassland, and the biomass has a higher amount of lignocellulose and a lower amount of nutrients that makes them more advantageous for renewable energy supply.

Against this background, it is the aim of this paper to evaluate the changes in land use from grassland to short rotation coppice (SRC) by reference to the Integrative Sustainability Concept. The focus is on SRC because fast-growing trees could contribute to the national wood resources and help to release the pressure on forests to become more productive in satisfying the growing demand for wood as an energy carrier which has been doubled in Germany within the last 5 years. The reserves to expand the use of wood in public forests without impairing sustainable and multifunctional forestry are limited due to ecological constraints. In small-size private forests with still higher wood reserves, structural problems prevent the mobilization of residual forest wood [4].

In Germany and also in other European countries, the cultivation of SRC has been intensively discussed and explored. According to the EU premium rights and to the German Federal Forest Act, SRC set up on agricultural land do not fall within the definition of forest. Hence, SRC maintain the legal status of agricultural areas as long as the rotation period is not longer than 20 years. Meanwhile, in Germany, around 4,000 ha of SRC have been established mainly for research and demonstration purposes. Energy scenarios are assuming that large agricultural areas are usable for SRC [5,6]. The European Environment Agency estimates that between 824,000 ha today and around 3.4 million ha in 2030 could be available for biomass crop production in the EU-15 (Austria, Belgium, Denmark, Finland, France, Germany, Greece, Ireland, Italy, Luxembourg, the Netherlands, Portugal, Spain, Sweden and the United Kingdom). More than half of which is available as former grassland[7].

METHODS

The Integrative Sustainability Concept [8] provides a theoretically well-founded approach to operationalize the *Leitbild* of sustainable development and an operable analytical tool for sustainability analyses. It has been applied so far successfully in various research projects on the global, national and regional levels [9]. The substantial minimum requirements of the Integrated Sustainability Concept (see Table 1) were chosen to identify relevant criteria for the assessment of the change in land use from grassland to SRC.

Table 1: Objectives of the integrative sustainability concept[8]

Securing mankind's existence	**Upholding society's productive potential**	**Keeping options for development and action open**
Protection of human health	Sustainable use of renewable resources	Equal access to education, information, and an occupation
Securing the satisfaction of basic needs	Sustainable use of non-renewable resources	Participation in societal decision-making processes
Autonomous self support	Sustainable use of the environment as a sink	Conservation of the cultural heritage and of cultural diversity
Just distribution of chances for using natural resources	Avoidance of unacceptable technical risks	Conservation of nature's cultural functions
Compensation of extreme differences in income and wealth	Sustainable development of real, human, and knowledge capital	Conservation of social resources

Rösch *et al.*

Rösch *et al. Energy, Sustainability and Society* 2013 3:7, doi:10.1186/2192-0567-3-7

The adaptation of the Integrative Sustainability Concept to the context of land use changes requires the identification of the relevant objectives as well as the selection of appropriate sustainability goals and indicators to measure the progress in reaching these goals. The sustainability objectives which have been considered to be most relevant in the context of land use change to SRC are listed in Table 2.

Table 2: Sustainability objectives for changes in grassland use

Sustainability objectives	Explanation
Securing the satisfaction of basic human needs	Affordable provision with basic supplies such as food and energy and prevention of competition between the use of commodities
Autonomous self support	Adequate and stable economic existence by income-producing employment through agricultural land use
Sustainable use of renewable resources	Warranty of the efficiency, reliability and replacement of renewable resources
Sustainable use of non-renewable resources	Substitution of non-renewable fossil energy carriers with bioenergy
Sustainable development of real, human, and knowledge capital	Priority for feedstock production to operate existing food processing and bioenergy plants
Conservation of nature's cultural functions	Preservation of cultural and natural landscapes of particular characteristic and beauty

Rösch *et al.*

Rösch *et al. Energy, Sustainability and Society* 2013 3:7, doi:10.1186/2192-0567-3-7

The conversion of grassland to SRC affects the objective to secure the supply of food as a basic human need because grassland is the main feedstock for milk and cattle meat production. The ongoing grassland conversion and expanding urbanization contribute to the continuous loss of this food production resource. At the same time, transforming grassland to high-yielding SRC production sites is detrimental to food supply. There is disagreement about how significant the 'food versus fuel' debate is, but consensus exists that biomass use for food should be prior to energy production.

Farmers provide a strong economic and social base for rural development. Diversification of income sources such as the increasing involvement of farmers in renewable energy production (e.g. biogas plants) helps to ensure the autonomous self support of people working in the agricultural sector.

Agriculture has strong impacts on the use of the renewable resources soil, water and biodiversity. For example, agricultural practices are the single greatest contributor to the increase in soil erosion rates. Increasing production of renewable energy from biomass contributes to the substitution of fossil fuels and the avoidance of carbon emissions and climate change.

Agricultural activities rely on different types of capital including the real (natural) capital - the soil resource, water from rainfall or other sources, the air, animals used for their labour and as a source of manure, the surrounding natural vegetation; the human capital - humans who supply labour, not only physical labour but also intellectual input for planning production strategies; the social capital - systems providing labour and marketing support as well as information related to agriculture and health services; the financial capital - markets for purchase and sale of goods and a credit system supplying funds to all levels of agricultural workers; and human-made capital - implements needed for agriculture, roads and means of transport and factories for processing of farm produce. At every level, an agricultural system depends 'on the value of services flowing from the total stock of' these five types of capital. (Table 2).

The choice of objectives is accomplished by scientific expertise and based on reflections with regard to criteria relevance, diversity, comprehensibility and data availability. Consequently, the set of criteria is neither complete nor perfectly consistent with the initial proposition nor fully developed in terms of the complexities of systemic interactions and cross impacts. The preliminary character of the set of sustainability criteria proposed in this paper is amplified by the fact that an agreement has to be brought about regarding the selection of sustainability criteria preferably via negotiations within society and politics. Such a transdisciplinary approach, however, would have gone beyond the scope of the research described in this paper. In the following section, the characteristic aspects of the sustainability objectives listed in Table 2 will be discussed.

Satisfaction of Basic Human Needs

The ongoing increase in agricultural yield has first led to overproduction of food in Germany and the EU and then to the implementation of set-aside programmes to restrict food production. The subsequent set-aside areas have been unlocked from the production ban for the purpose to grow crops for material or energy supply. In Germany, now more than 2.15 million ha or 18% of the agricultural land are used for the production of renewable resources. Growth of crop yields has slowed down considerably, and fears are expressed that the trend may not reverse [10]. The question is whether the lower growth potential is sufficient to meet the increased requirements. Climate change, furthermore, looms largely as a risk that would negatively affect the production potentials of agricultural resources.

The increased request for agricultural land by the concurrent production of food and energy is visible in rising prices for the purchase and lease of agricultural land respectively in the neighbourhood of biogas plants [11]. The price lift for arable land is much higher than for grassland, indicating the restricted land use options for grassland compared to arable land. The boosters of the advancing demand for land in Germany are the economic incentives given by energy politics in order to reach the ambitious renewable energy goals.

Food production should be the prior land use because renewable energy can be supplied by other types of renewable resources, such as wind, sun, water or geothermal power. The human need for food however can exclusively be covered by biomass produced on land and water-based production systems. Concerning this ethical aspect of land use, grassland needed for livestock grazing respectively for cattle farming or feeding other ruminants, such as sheep, should not be used for establishing non-food crops such as SRC.

The question though is how much grassland is needed to accomplish food supply today and in the future in a world based on the division of labour and resource use. In Germany, grassland is characterized by a long-term decline without visible impact on the food supply. Between 1990 and 2006, the area of permanent grassland in Germany declined at an annual rate of 0.8%, while arable land only decreased at a rate of 0.05% per year [1]. The overall loss of farmland due to urbanization seems to have occurred mainly at the expense of grassland [12]. This development is the result of the decrease in dairy cows due to political and economic constraints (i.e. the imposition of milk quotas) and the progress in breeding and milk production which have subjected milk production to significant structural changes for many years [13]. The average annual milk production per cow in Germany has increased from 4,710 kg in 1990 to 7,080 kg in 2010 [14]. Similarly, there has been a reduction in the number of beef cattle following the Bovine spongiform encephalopathy crisis in the late 1990s and the reduction and eventual removal of specific beef production incentives. As progress in breeding and dairy cow milk production will continue, a further decline in the demand of grassland can be expected.

Besides progress in breeding and milk production, the demand of grassland is influenced by the spatial allocation of the dairy and cattle stock and also by the intensity of dairy cow milk production and feed production on grassland and arable land. These factors are strongly dependent on economics, which in turn is influenced by agricultural policies and incentives. The objective of the German Sustainability Strategy to increase the share of extensive organic farming from 5.9% in 2010 to 20% will increase the demand of grassland for roughage production [15].

Since grassland is just one source of animal feed, its competiveness as compared to maize is crucial for its application [13]. Until the 1970s, grassland, lucerne and clover had been the main feeding base for cattle farming. Today, these traditional forage plants have been replaced continuously by maize. This is due to the area-specific higher yield of maize in contrast to grassland. This configuration will not be changed in the future as long as the spread of pests such as the corn rootworm will not limit maize production.

The conclusion is that grassland can be replaced by crops with higher yields per hectare such as maize. However, importing feed like soy bean from Latin American countries instead of using domestic grassland for the supply of animal feed raises ethical issues.

Autonomous Self Support

In common with most high-income countries, the agriculture's contribution to the German gross domestic product and employment is low, although agriculture is the largest type of land use accounting for about 52% of the total area. The regional importance of the sector and its associated food and farming industries varies considerably. The job and income opportunities associated with agricultural land use are strongly influenced by government interventions in agricultural markets. They are likely to be the most critical drivers of agricultural land use change with significant influence on the landscape.

The German agricultural policy has primarily been determined by the European CAP covering the key agricultural commodities. The CAP was devised to increase agricultural productivity, stabilize markets and assure the availability of reasonably priced food, and to ensure a fair standard of living for all those involved in farming. These objectives were largely met by the support of internal prices and incomes, through direct market intervention, and border protection through tariffs and levies on food imports. Export subsidies were paid to offload otherwise uncompetitive EU surplus production onto world markets. The policy was successful in terms of increasing domestic food production, to the extent that the EU became the second largest exporter of food in the world. However, this policy resulted in higher consumer food prices within the EU and increasing of tax burdens on EU citizens. Its legitimacy as a means of supporting farm and rural incomes was eventually challenged by food exporting nations and the World Trade Organization.

The drive for food production between 1940 and 1984 had led to substantial areas of permanent pasture being ploughed and drained, usually with grant support. By 1984, surplus milk production led to the introduction of marketable milk quotas. Eight years later, in 1992, the MacSharry reforms of the CAP led to constraints being placed on the beef and sheep sector, with compensation payments which were subject to regional ceilings and maximum stocking rates. At the same time, compensation payments in the arable crop sector required the compulsory withdrawal (or setting-aside) of up to 10% of arable land from production.

The negative environmental effects of intensive farming and land use change including loss of habitats and wildlife, soil erosion and water pollution have lead to a change in agricultural politics. Today, payment for farmers is not linked anymore to a particular type of production, but does require 'cross-compliance', an adherence to EU environmental, food safety, animal welfare standards and regulations, and the need to keep farmland in good agricultural and environmental condition. This is defined at a national level and includes the maintenance of grassland.

For grassland different incentives are available on the national and regional level, but for SRC, the payments are limited. With respect to the on-farm economic competitiveness of SRC toward the classical use of agricultural land under German conditions, it might be profitable to establish SRC, but today this is mainly not the case [15-17]. The establishment of SRC is hampered by traditionalistic behaviour and risk aversion, too. SRC involve investment costs that are at least partly sunk or irreversible, and they are changing the farm's revenues and expenditures over a long period of time. Farmers would go for SRC if the present value of the investment returns exceeds the investment costs considerably or additional and long-lasting area payments for SRC are provided by policy and society [18]. Farmers do not want to allocate their land for long periods of time and rather prefer great flexibility in their planting decisions in order to rapidly react to changes in prices for agricultural commodities.

Sustainable Use of Renewable Resources

The EU has acknowledged the various ecological functions of grassland for the preservation of renewable resources and its benefits in terms of soil, water and biodiversity protection. Hence, the EU has introduced measures within the CAP for its protection against conversion into arable land as well as for preserving or enhancing its ecological quality. Extensively cultivated grassland belongs to the most species-rich land use types in Europe [19] and, after being destroyed, may only recover very slowly [20]. The conversion of this type of grassland could lead to a loss of biodiversity and should thus be excluded from changes in land use.

In recent times, grassland has received considerable attention for its carbon sink capacity and high carbon stock compared to arable land [21]. In particular, grassland on organic soils store very high amounts of carbon [22,23]. The conversion from permanent grassland to arable land is accompanied by a massive boost of mineralization due to decomposition of soil organic matter and consequently rising greenhouse gas (GHG) emissions as well as higher nutrient runoff and leaching to surface and ground water. Converted grassland may lose about 50% or more of its original soil carbon under arable management [24]. In the case of temperate grasslands which contain in average around 330 mg of carbon per hectare, this is a tremendous release of GHG [25]. In contrast, a conversion of arable land to grassland results in increased carbon storage, however, at a much slower pace and rarely reaching the former level. Compared to annual agricultural crops such as rapeseed or maize, fast-growing trees have significantly lower negative impacts on the environment [24,27-29]. This is related to the extensive form of cultivation which can lead to an increase of carbon and humus in the soil and their low demand for nutrients and pesticides [30,31]. Established SRC have a high economic threshold to pest damage and can be managed with few pesticide applications. Tolerating the presence of certain insects and other plants is, in turn, cited as the main reason why SRC, if appropriately located, has the potential to increase biodiversity in many farmland situations [32].

The establishment of SRC on grassland is less valuable for the protection of resources and the environment than on arable land. The deterioration of the ecological functions of grassland can be reduced by the non-plough tillage, minimal invasive establishment of SRC and the abandonment of returning the SRC plantation into grassland or arable land [33]. Hence, SRC can contribute to both the buildup of humus and improvement of the soil quality, in particular, on soils with low carbon stocks [34]. This assumption is eligible because fast-growing trees in contrast to annual non-woody energy crops do not require good carbon-rich soils to achieve economically promising yields. They can also be produced on marginal agricultural sites with a low natural productivity, like sandy or stony soils; if the average temperature enables long vegetation periods, the requirements by the species are respected and water supply is guaranteed due to access to groundwater resources or sufficient precipitation [30].

SRC can also be beneficial for the protection of biodiversity within sparsely wooded pastoral landscapes where they could provide wooded stepping stones for birds of wooded habitats [35,36] and create networks of biotopes and enhance biodiversity [37,38]. Positive effects for biodiversity can also be expected when SRC are established in riparian buffer stripes to ensure that the regulations of distance regarding the application of fertilizer and pesticides will be respected [30] and in emission stripes along transportation routes to shield the fields behind against pollutants [31]. In water regulation areas, SRC can contribute to retard peak flows, decrease surface runoff and utilize the soil as a buffer [28,30]. Against this background, it can be concluded that even on grassland the establishment of SRC can be beneficial for the preservation of renewable resources and the environment. The areas where this is true and which are not reaching a certain critical value of natural protection and the preservation of ecosystem services [27,28], should be identified and made available for establishing SRC. The measure for planting and operating SRC has to be conducted environmentally sound and needs to fit into the general principles for saving both natural and cultural landscape [27-29,39].

Sustainable Use of Non-Renewable Resources

Energy and carbon analysis of different power generation processes based on SRC demonstrated that these processes use comparatively less fossil fuel and avoid more greenhouse gas emission per unit of produced energy than conventional high-input bioenergy crops [40]. The energy ratio of SRC is within the range of 22 and 28 depending on both the assumptions made with regard to the production and transformation processes and on the fact which energy consumptions along the process chain are included [41,42]. SRC are well positioned among the best energy crops and can contribute to increase the share of bioenergy compared to the energetic use of grassland and thus the amount of fossil energy carriers that can be substituted.

Sustainable Development of Real, Human and Knowledge Capital

In times of financial crises, the sustainable use of capital is of increasing importance for a sustainable development. Investments in bioenergy plants, such as biogas plants, which have been made in the last years based on the fixed feed-in tariffs specified by the German Renewable Energy Sources Act (EEG) could be endangered by rising prices for biomass feedstock due to an increasing competition for land use. Among the means of using grassland biomass, biogas production currently is the most common practice in Germany. Surveys of agricultural biogas plants in Germany show that grass silage is used as a feedstock in more than 50% of the biogas plants and is the second most frequent crop feedstock after maize silage [43]. Particularly, in regions with high percentages of grassland, the relevance of grass silage as a biogas feedstock could be quite high. To deal with the capital as a limited financial resource in a sustainable way, it is defined in this paper that grassland used to produce grass silage for already existing or planned biogas plants should not be taken into consideration for the establishment of SRC.

Conservation of Nature's Cultural Functions

The functions of nature are mainly associated with its life-support and ecological services. The important and irreplaceable cultural functions which are performed by natural environment are rarely investigated. This is due to the challenge to determine indicators and values for these sociocultural functions and their values for the health and well-being of human societies. The importance of the information functions (health, recreation, amenity, education, heritage, etc.) for the quality and sustainability of human life, however, is undisputed. Despite their immaterial and often intangible nature, these functions provide many socioeconomic benefits which might be assessed through both qualitative and quantitative evaluation methodologies. Depending on the spatial embedding and the design of SRC plantation, they can either enrich or depreciate the cultural landscape. The impact on the landscape largely depends on the suitability of the proposed location and on the current character and sensitivity of the landscape. In general terms, lowland landscapes with high levels of tree and woodland cover and arable or mixed farming are most suitable for SRC. It is also important to remember that cropping is heavily mechanized and requires land suitable for mechanical operations: This excludes steep or boggy ground.

The impact of SRC on the landscape will increase if whole farms or substantial parts of farms are planted. There is a risk that the diversity of the landscape will be reduced and that it will appear saturated [44]. It is important to build diversity into large-scale planting by varying age structure and introducing open space so that the crop is subdivided at a scale that suits the particular landscape type. Some landscapes, such as parkland and historically designed landscapes may be unsuitable for the introduction of SRC on any significant scale. In areas well used by the public such as near public rights of way, access routes and roadsides, a more detailed design may be necessary to respect cultural functions.

Indicators for Sustainable Grassland Use Change

The sustainability targets can be operationalized by the application of a readily comprehensible set of sustainability indicators. This way, the quality of the dialogue about sustainable land use change from grassland to SRC can be enhanced. The selection of suitable indicators is based on the following criteria: relevance, directional safety, availability of data, comprehensibility, and practicability. The indicators chosen to identify potential grassland areas which could be used for a sustainable establishment of SRC are listed in Table 3.

Table 3: Sustainability indicators for identifying grassland suitable for SRC

Sustainability indicators for identifying grassland suitable for SRC	
Sustainability objective	**Sustainability indicator**
Securing the satisfaction of basic human needs	Grassland no longer needed for animal feed production
Sustainable use of renewable resources	Grassland with suitable natural conditions with regard to temperature and water availability for fast growing trees
	Grassland with no particular function for the protection of biodiversity or ground water
Sustainable development of real, human, and knowledge capital	Grassland without feedstock functions for biogas plants
Autonomous self support	Employment and income by grassland use

Rösch et al.

Rösch et al. Energy, Sustainability and Society 2013 **3**:7 doi:10.1186/2192-0567-3-7

The analysis of sustainability indicators in this paper is focusing on the most important indicators regarding their relevance to illustrate the sustainability objectives which are strongly affected by the conversion of grassland to SRC and considering the availability of data. In the following, these indicators will be applied in the state of Baden-Wuerttemberg. Besides their purpose for sustainability assessment, they can be used to enhance communication and discussion in society and politics on options for sustainable land use changes. Moreover, they can contribute to provide opportunities for creativity in planning of how to achieve sustainable development and support decision-makers.

RESULTS

As a case of application of the selected sustainability indicators the federal state of Baden-Wuerttemberg in Germany has been chosen due to its still high percentage of grassland and the fact that its cattle stock is continuously decreasing. The pattern of grassland use is spatially diverse because of the various natural conditions influenced by different combinations of climate, soils and topography. Major grassland areas can be found in the region of the Black Forest and the Swabian Alb where agricultural land use is difficult due to sparse soils, slopes, temperature and high precipitation.

Grassland Suitable for SRC

Suitable temperature levels and the sufficient availability of water are crucial cultivation factors, especially for poplar and willow grown as SRC [45,46]. The water availability can be described by the relief-dependent (influenced by the exposition, slope and relief position) climatic water balance within the period from May to October and the available water capacity in the upper 1 m of the ground. On 62 polar and willow plantations in south-west Germany and in the northern part of France, the water availability in millimetre and the yields in $tons_{dm}$ after the first three-year rotation were measured. On the basis of this data, a correlation between the water availability and the achievable yields could be derived and expressed in an equation [47]. The water availability was classified in five water availability stages (<0, 0 to 100, 100 to 200, 200 to 300 and >300 mm) to better visualize the natural SRC cultivation potential in maps. These classes represent the suitability of a site for the cultivation of poplar or willow grown as SRC. The classes are sorted in ascending order from <0 to >300 mm in these categories: excluded, unfavourable, medium, favourable and very favourable. The achievable yields on wet and waterlogged sites were also measured and classified as favourable and unfavourable by a mean annual temperature of more than 8°C. To consider the influence of the temperature on the growth of SRC, the mean annual temperature was classified in three classes: <6.5°C, 6.5 to 8.0°C, and >8.0°C; and the SRC site classes were downgraded according to the tempe- rature class (see Table 4).

Table 4: Classification of location suitability for short rotation coppice according to water availability and mean annual temperature

Ø Annual temperature (°C)	Water availability stages						
	<0 mm	**0-100 mm**	**100-200 mm**	**200-300 mm**	**>300 mm**	**Wet**	**Water logged**
<6.5	Excluded	Excluded	Excluded	Excluded	Excluded	Excluded	Excluded
6.5-8.0	Excluded	Excluded	Unfavourable	Medium	Favourable	Medium	Excluded
>8.0	Excluded	Unfavourable	Medium	Favourable	Very favourable	Favourable	Unfavourable

Rösch et al.

Rösch et al. Energy, Sustainability and Society 2013 3:7 doi:10.1186/2192-0567-3-7

Since average temperature and water availability have a strong influence on the land use efficiency and economic performance of SRC, these criteria have particularly been chosen to identify land which is suitable for the establishment of SRC in the research region Baden-Wuerttemberg. In applying these criteria, it can be concluded that around 12% (65,000 ha) of the grassland in Baden-Wuerttemberg shows very favourable and around 30% (160,000 ha) favourable conditions for the cultivation of SRC. In total, over 40% of the grassland can be rated as suitable for fast-growing poplars and willows (see Figure 1).

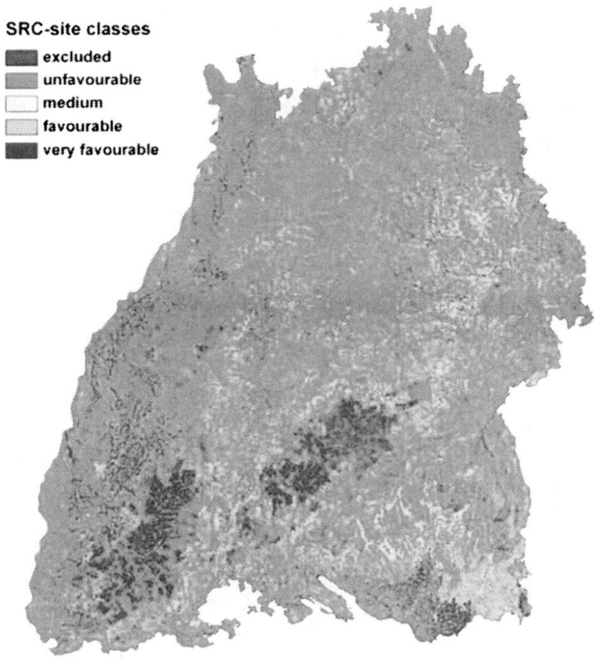

Figure 1: Suitability of grassland area in Baden-Wuerttemberg for SRC with regard to site conditions[47].

The percentage of suitable land for poplar and willow SRC within the respective municipalities is considerably high in the south-east of Baden-Wuerttemberg (see Figure 1). Also, parts of the north-eastern mountain regions and the area between the Black Forest and the Upper Rhine Valley possess considerable land potentials.

Grassland without Feeding Functions

In Baden-Wuerttemberg, the number of cattle has decreased from around 1.9 million cattle in 1970 to around 1 million cattle in 2010. In the same period, the number of dairy cows has declined from 750,000 to 350,000. Compared to the German average in annual milk production per cow of around 7,080 kg, the average milk in Baden-Wuerttemberg is significantly lower at 6,579 kg [14]. Thus a further decline in the number of dairy cows can be expected due to progress in milk production and structural changes [2]. This decline is associated with a release of grassland not needed anymore for the production of cattle feed. Part of this so-called surplus grassland has been converted in the last years to arable land in order to produce maize. Agricultural policies and structural changes have shifted the milk production to regions with favourable natural production conditions. The nationwide trade of milk quota has further decreased the number of dairy cows in Baden-Wuerttemberg. Within 3 years, almost 5% of the total milk production has been translocated to the other states [48]. This correlates with a dismount of 15,000 cows or a surplus area of 7,000 ha grassland.

For the calculation of the surplus grassland, a method was developed to assess and allocate the surplus grassland on the basis of different statistical data on land use, livestock and agricultural yields in Baden-Württemberg as described in [49]. Further data needed for the calculation such as the energy content of grass silages were raised by a special two-step survey among the local agricultural offices. First, based on these data, the roughage demand of cattle, horses and sheep was calculated. Then the production volume of the different types of grassland (meadows, pasture) as well as other fodder crops was calculated in relation to the average yield and energy content on district level. Finally, the production volume was balanced with the estimated roughage demand of the livestock. For more information, see [49]. According to this method for calculating the amount of surplus grassland, it has been assessed that in Baden-Wuerttemberg, currently around 100,000 ha of the permanent grassland are not needed anymore for roughage production. The spatial distribution of this surplus grassland is quite different (see Figure 2). The main part of this surplus grassland which could be used for establishing SRC is located in the region of the Black Forest and the Swabian Alb.

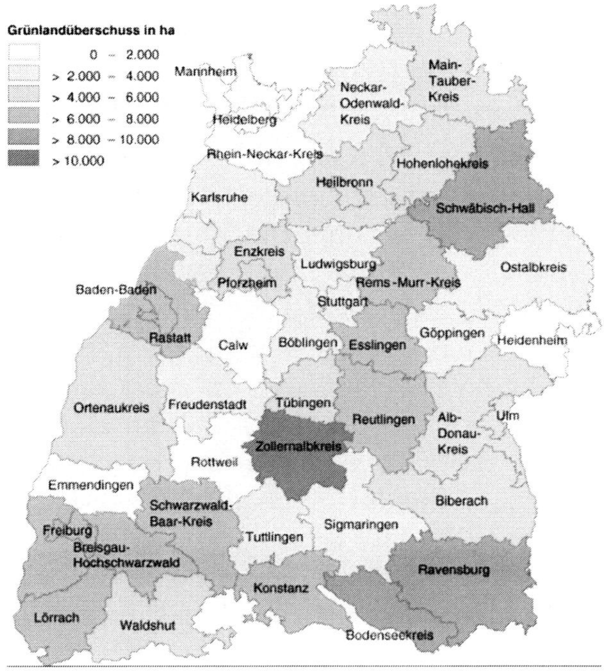

Figure 2: Grassland without a function for cattle feeding in Baden-Wuerttemberg.

At present, the exploitation of the potential of this grassland for SRC is limited by political frame conditions. Under the EU Cross Compliance Regulations within the Common Agricultural Policy, EU member states are required to ensure that land under permanent pasture is to be maintained. Under present EU legislation, there is a general derogation which allows some decreases provided that they are not significant - defining significant as losses which exceed 10% of a reference level - to be set by the member states themselves. In Germany, this level was set to 5% of the reference level to the correlation between grassland and arable land in 2003. In many states of Germany, the limit of 5% has already been reached. In the research region, only 24,000 ha or 2.38% of the referring grassland has been converted so far. Thus grassland could still be converted and used for fast-growing trees. However, legislation in Baden-Wuerttemberg has changed. Since December 2011, the conversion of grassland is prohibited mainly to prevent that more grassland is turned to maize fields for biogas plants. The establishment of SRC on grassland into SRC plantations is also generally excluded by law [50].

Grassland without Function for Biodiversity or Groundwater Quality

Environmental national and regional legislation in the areas of nature conservation and water and soil protection may limit the conversion of grassland in ecologically sensitive locations, such as nature protection areas and drinking water protection areas or flood plains, supporting both national and EU objectives. Grassland possessing a high variety of species is protected respectively by the FFH-guidelines [51], a specific EU legislation that classifies these areas as 'flora fauna habitats' (NATURA 2000 areas). NATURA 2000 is an ecological network of protected areas in the territory of the European Union based on the Habitats Directive and the Birds Directive, and protects around 18% of land in the EU countries. Hot spots of distribution for this type of land use in Baden-Wuerttemberg can be found in the mountain areas of the Swabian Alb and, more often, in the Black Forest. Grassland is also protected by the EU guidelines for bird protection, which correspond to 17.3% of the area of Baden-Wuerttemberg or about 620,000 ha of the NATURA 2000 areas. Amongst them, the proportion of grassland is about 110,000 ha, plus another 30,000 ha of fruit meadows. The protection status of permanent grassland in NATURA 2000 areas, being equivalent to 15.4% of the German land area, is not clear. In cases where they do not overlap with nature conservation areas, protected biotopes or other areas with defined restrictions on land management, grassland may be converted if a compatibility assessment is conducted featuring that no significant changes in land use are induced. Besides, land use changes are possible in these protection areas as long as this does not result in a significant deterioration or disturbance in relation to the protected habitats or species and it is ensured that habitat conditions would not be affected negatively [1]. This requirement is somewhat vague and apparently leaves some scope for conversion of grassland into arable land, particularly in the case of special protection areas designated under the Birds Directive.

About 25% of the grassland in Baden-Wuerttemberg is classified as 'extensively cultivated grassland' rich in species. In 8% of this area, the biodiversity is judged as very high [52] whereas 1% is classified to be a 'biotope grassland'. The agri-environmental programme of Baden-Wuerttemberg known as 'Marktentlastungs-

und Kulturlandschaftsausgleich' or MEKA II [53] acknowledged about 70,000 ha or 12% of the grassland areas as rich with species. This is just about half of the total extensively cultivated grassland classified this way. High amounts of this species-rich grassland are situated in the Swabian Mountains' districts of Zollernalb and Tuebingen. The preservation of biodiversity is regarded as superior sustainability goal than the production of renewable energy because this objective can be achieved by solar and wind energy too with less impact on biodiversity. Thus, the following categories of grassland are excluded from the establishment of SRC on grassland:

- Areas for the preservation of the diversity of species, officially nominated as Flora Fauna Habitats or Bird Protection Areas (NATURA 2000)
- Natural conservation areas, national parks, core zones of biosphere reserves, protected biotopes according to Section 30 of Federal Nature Conservation Act or BNatSchG,
- Permanent grassland cultivated extensively.

The high water demand of fast-growing poplars and willows can have an influence on the water availability in groundwater bodies. Thus, land areas with susceptible ground water regimes and crucial functions for water quality and functionality of water bodies are not regarded as suitable locations for sustainable cultivation of fast-growing trees. Figure 3 illustrates the spatial distribution of the areas for the protection of nature, landscape, biodiversity and water in Baden-Wuerttemberg.

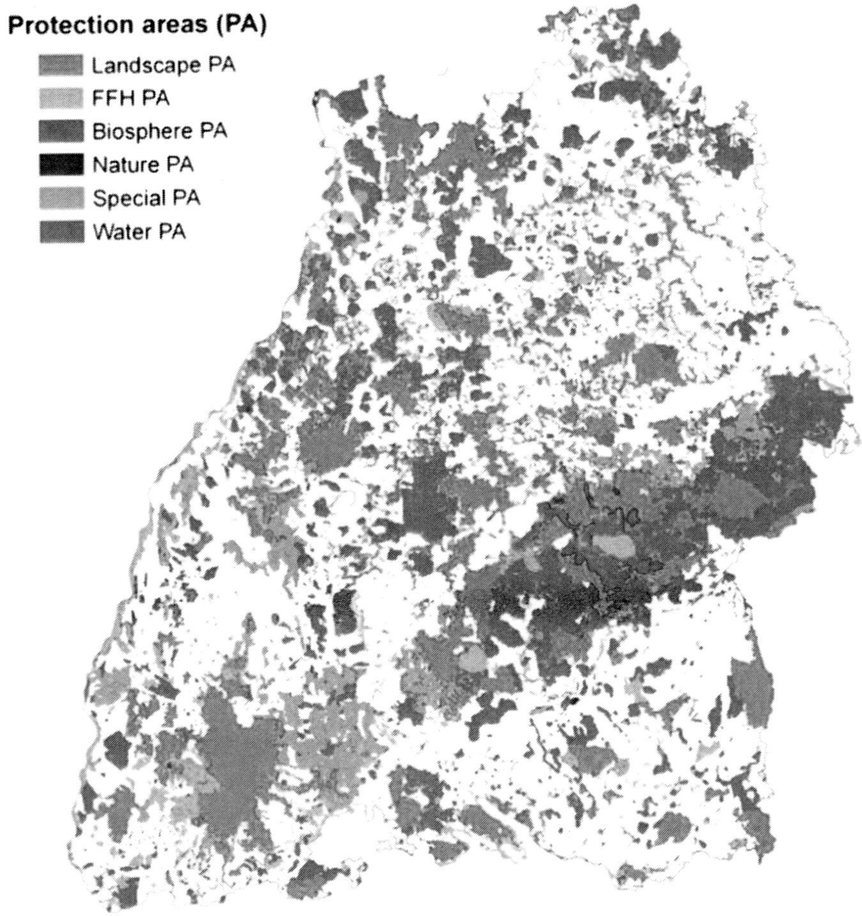

Figure 3: Spatial distribution of the areas for protection of nature, landscape, biodiversity and water in Baden-Wuerttemberg[47].

Figure 4 illustrates the areas on which the establishment of SRC is not possible or restricted and the area where fast-growing trees could be planted without impairment of biodiversity and water resources.

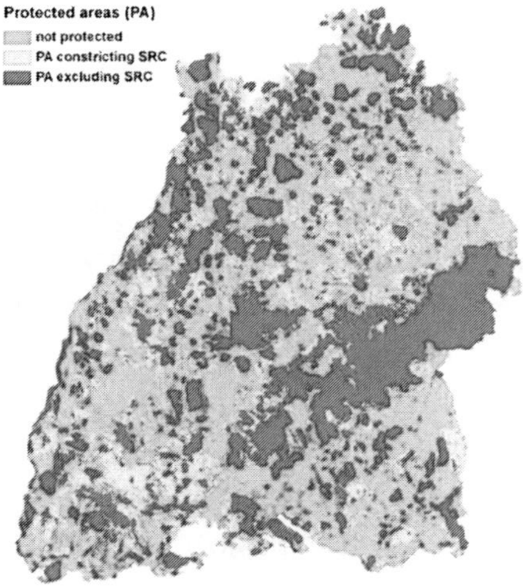

Figure 4: Locations where the cultivation of SRC is restricted due to environmental aspects[47].

Grassland Without Biogas Feedstock Functions

Grassland provides grass silage which is a good feedstock for biogas plants that can easily be stored. Grass silage is applied in over 50% of the biogas plants and represents the second most important substrate type on a federal level providing 11% of the whole amount of biomass for biogas plants and allocating 12% of permanent grassland [43]. In Baden-Wuerttemberg, the average proportion of meadow grass differs between 9% in large plants >500 kW and 15% in small plants of 150 kW [54]. The distribution of biogas plants and the proportion of grass used for biogas plants are spatially different as shown in Figure 5. This is not only the influence of the spatial distribution of small and large-scale biogas plants but also due to the relation of grassland to arable land which is quite unequal in the various districts. In the south of the Black Forest (mainly in the districts Breisgau-Hochschwarzwald and Tuttlingen), grassland is the main source of feedstock for biogas production delivering more than a third of the biogas feedstock. In Baden-Wuerttemberg, the installed capacity of biogas plants has continuously increased until 2011 up to 256 MW of electricity. Assuming that in average 11% of the biogas feedstock supply for these plants derives from grassland, about 27,000 ha of grassland are needed to produce substrate for biogas plants. The increased use of grassland as a biogas feedstock resource could be one way to counter the widely expressed criticism of the significant expansion in maize cultivation for biogas plants. Despite its ecological advantages compared to maize, an enlargement of the share of grass feedstock is not assumed here because permanent grassland is less competitive in many regions compared to cropping systems [55]. The preference for growing silage maize for biogas production is based primarily on its high yields per hectare. In order to facilitate the use of grass, the Renewable Energy Act (EEG) 2012 introduces upper substrate limits for maize and cereal grain (ethical discussion) and different raw material payment classes (0, I, II). The intention is to expand the cultivation and use of grass and alternative substrates by providing a higher payment for the feedstocks.

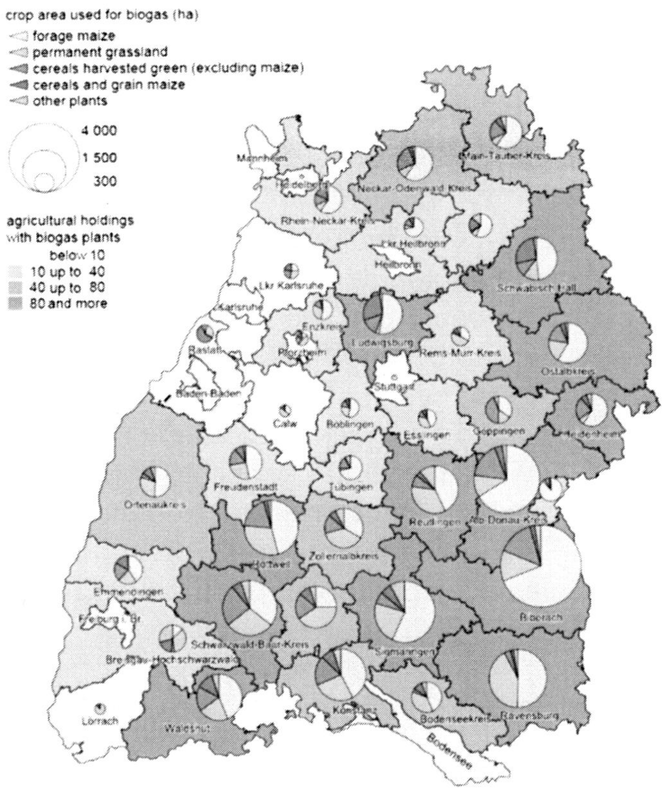

Figure 5: Grassland used for biogas production in the districts of Baden-Wuerttemberg[55].

This area of grassland assigned for biogas production is not considered to be disposable for establishing fast-growing trees in order to avoid feedstock competition. This prioritization is based on the sustainability goal not to jeopardize the sustainable use of real and human capital which has already been invested in local biogas plants and operation. Besides, the total area occupancy caused by biogas plants is already quite high in districts with large amounts of cattle livestock. Thus, negative impacts on the availability of land and land tenures have already been documented for such areas [56,57]. Finally, only grassland which is not dedicated to biogas production should be considered for the establishment of SRC.

Employment and Income by SRC on Grassland

Milk and meat production based on dairy cows and cattle is the main source of employment and income from grassland. In Baden-Wuerttemberg, one quarter of the agricultural production value is generated on grassland [58]. As usual, the spatial distribution of this production value is differentiated. The main production areas are located in regions with favourable natural conditions for milk production such as the Allgaeu, Upper Swabia and the region from the Ostalb to Hohenlohe (see Figure 6). Over 30% of the total milk production is generated by two districts in the south-eastern part only, Ravensburg (464,100 t/a) and Biberach (226,000 t/a). However, also in these districts, grassland which is no longer used for milk production could be used for the establishment of SRC. The land in question is mainly marginal grassland with regard to the practicability for cultivation with large-sized efficient harvest machinery [2]. The establishment of SRC will result in a reduction of employment and income compared to grassland which is used as resource for milk and meat production due to afar less labour-intensive production chain. Yet, if SRC are grown on surplus grassland without functioning as dairy and cattle or biogas feedstock, a positive impact on employment and income for the local community can be realized. Hence, from the view of sustainable development, only the grassland area identified in the preceding chapters should be applied for the establishment of SRC.

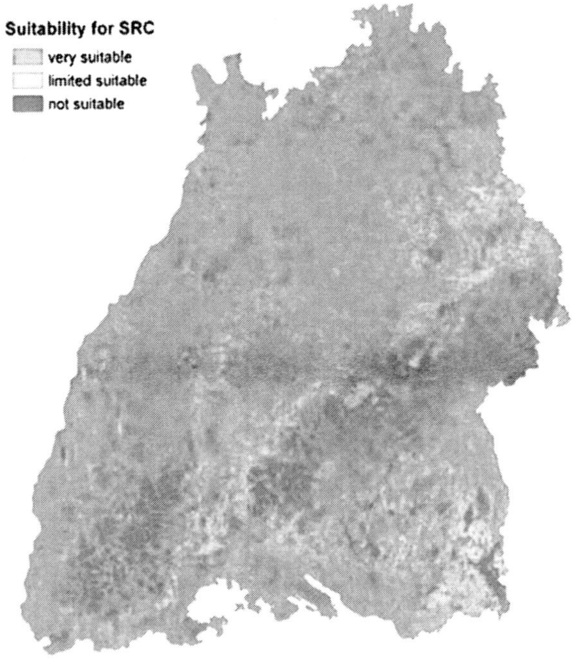

Figure 6: Suitability of grassland for the cultivation of SRC in Baden-Wuerttemberg with regard to sustainability criteria.

Grassland Suitable for SRC With Regard to Sustainability

The results achieved by the application of the previous sustainability criteria exhibit that in the research region, around 13% or 70,000 ha of grassland are suitable for the establishment of SRC. A main part of this grassland can be found in the south-eastern part of Baden-Wuerttemberg (see Figure 6).

DISCUSSION

The sustainability debate on land use change is often narrowed and reduced to the questions of compatibility with environmental and climatic conditions. However, land use change has to be evaluated against a much larger spectrum of sustainability criteria, including aspects of security of supply and economic development of rural areas. The adoption of the Integrative Sustainability Concept to analyse land use change from grassland to SRC has the big advantage that socio-economic and ecological aspects are regarded together and of equal rank. From the 15 principles of the Integrative Sustainability Concept, six principles have been regarded as relevant to describe the impacts of land use change. Aside from the sustainable use of renewable resources, the criteria referring to the satisfaction of basic human needs, the autonomous self support of the population and the development of real, human and knowledge capital have been identified as highly relevant for identifying grassland which could be transformed to SRC in a sustainable manner. Even if these restrictions are taken seriously, there are considerable areas of grassland available for the establishment of SRC as turn outs by the application of the sustainability principles in the state of Baden-Wuerttemberg.

The impacts on the cultural landscape heritage and nature's cultural functions, respectively, with regard to tourism and recreation are also crucial for assessing the sustainability and also the acceptability of land use change. Empirical surveys are needed to investigate the perception of land use changes from grassland to SRC on the levels of individuals, stakeholders, main actors and politicians, and group modelling is required to learn about the arguments and feelings behind the willingness for or resistance against land use change. In doing so, the preference of rejection of SRC has to be raised in comparison to other land use change for decentralized biomass and energy production technologies like open space photovoltaic or wind power. However, this would have been beyond the scope of this work. From lessons learned about land use change due to biogas plants and their demand of feedstock, it can be assumed that in particular large-scale SRC plantations could disrupt the character of the landscape and lead to the denegation of SRC. Hence, it will be of great importance to design and implement SRC in a way that the cultural and recreational landscape as well as human landscape perception, cognition and values are not affected negatively.

Regardless of the potential for sustainable land use change from grassland to SRC, the legislation in the state of Baden-Wuerttemberg prohibits the conversion of grassland at all. This regulation is a reaction to the conversion of grassland into arable land which begins with tillage, a practice known as negatively affecting soil properties and ecosystem processes in grassland-dominated ecosystems and to contributing to climate change by releasing soil-bound carbon stocks. On the EU level, the conversion of grassland is prohibited also but so far this rule is only applicable for land with high carbon stocks and high biodiversity which will be used afterwards for biofuel production according to the EU Renewable Energy Directive. These regulations are designed mainly for the protection of biodiversity; soil and climate provoke questions about the future of grassland which is and will be released from their traditional functions for roughage production and play no vital role for nature conservation purposes. An increase in the intensity of dairy farming and in biogas production as well as agricultural changes can lead to a continuing conversion of grassland, in cases when grassland is rented by other farms. Grassland without any production or nature conservation functions will no longer contribute to the farmer's income and the added value of the region but has to be maintained instead with public funds.

The ban on the change of grassland has not only consequences on local employment and income, but also on the decentralized supply and use of renewable and storable energy. Moreover, public funding of research on SRC in general and the development of methods to establish SRC are needed in a way that soil functions and carbon stocks are largely preserved. However, in other German states, such as the Saarland, the establishment of SRC even on grassland is subsidized by the government with up to 1,000 euros per hectare in order to combat climate change with local actions. The objective is to establish SRC on marginal land where competition from other crops is less severe, due to their lower economic return, and with less intensive cultivation systems closer to traditional forest management.

CONCLUSIONS

The potential for a sustainable change in the use of land from grassland to SRC in south-west of Germany is likely to be significant and can contribute to decentralized renewable energy supply, local employment, and income. The exploitation of this potential is currently hindered by the prohibition to transform grassland to other types of land use in Baden-Wuerttemberg. This decision is strongly influenced by ecological aspects and does not consider the chances provided by

SRC as an attractive alternative for a more extensive farming type and as additional local source for added value in economic and ecological terms. It is recommended to protect grassland more stringently on ecologically sensitive sites and also to strengthen legal standards of nature and water protection, but with regard to the lost economic values, not to set grassland conservation as an area-wide imperative.

In recognition of the advantages of SRC for sustainable land use, continuous research and demonstration of SRC are needed for improving the knowledge about site suitability and yields under current and future climates and frame conditions. Currently, only limited information is available for understanding the bottom-up spatial constraints for SRC production and both the changes in carbon stocks and the release of greenhouse gases, which are strongly dependent on the location, and the procedure how to establish SRC on grassland. More research is also required on supply chains to comprehend how demand and resource location affect costs and uptake in different end uses, and therefore inform on infrastructure and technologies that will be required.

Wood is the main renewable energy source in Germany and the government has ambitious targets to further increase the provision of bioenergy. Although all energy crops contribute to energy and climate-related sustainability goals, SRC are more favourable than annual energy crops to reach the targets. Nevertheless, the area of SRC in Germany is still very small and only slowly rising. This is mainly due to economic constraints and the lack of competiveness because of lower profit and return on investment compared with traditional crops. New and better incentives are needed for the establishment of SRC. Besides, more differentiated subsidies strategies based on spatial land use considerations have to be developed in order to achieve a better direct land use change. To avoid repeating the mistakes which have been made with other subsidies for energy crops and to ensure that the public money is achieving overall profit, sustainability criteria should be applied to allocate the SRC in locations where they can contribute best to nature and society.

By applying these criteria in the state of Baden-Wuerttemberg in a spatially differentiated bottom-up assessment, it was found that around 13% of the grassland might be available for SRC. Further research is needed not only in order to specify this preliminary allocation process and to tap the identified potential, but also in order to clarify whether

SRC should mainly be established on agricultural land which cannot be used in a highly productive way due to diverse local disadvantages such as along small rivers, very wet areas or small fields. Strategies and measures need to be developed and implemented to direct the establishment of SRC to the most suitable areas. Moreover, it is necessary to integrate stakeholders, main actors and the general public into the process of identifying areas for SRC. A major challenge will be to gain acceptance for the changes in landscape which will be evoked by establishing SRC, as they can lead to negative impact on the cultural and recreational functions of the landscape.

AUTHORS' CONTRIBUTIONS

CR formulated the concept and drafted the manuscript. CA participated in the spatial potential assessment. JJ contributed to the 'Methods' section. All authors have read and approved the final manuscript.

REFERENCES

1. Nitsch H, Osterburg B, Roggendorf W, Laggner B (2012) Cross compliance and the protection of grassland - illustrative analyses of land use transitions between permanent grassland and arable land in German regions. Land Use Policy 29(2):440-448
2. Rösch C, Skarka J, Raab K, Stelzer V (2009) Energy production from grassland - Assessing the sustainability of different process chains under German conditions. Biomass Bioenergy 33:689-700
3. Hartman A (2012) Dauergrünland in Baden-Württemberg. Statistisches Monatsheft Baden-Württemberg 2(2012):33-37
4. Mantau U, Saal U, Prins K, Steierer F, Lidner M, Verkerk H, Eggers J, Leek N, Odlenbuerger J, Asikainen A, Anttila P (2010) EUwood - real potential for changes in growth and use of EU forests. Hamburg, Germany: Final report. p 160
5. Fritsche UR, Dehoust G, Jenseit W, Hünecke K, Rausch L, Schüler D, Wiegmann K, Heinz A, Hiebel M, Ising M, Kabasci S, Unger C, Thrän D, Fröhlich N, Scholwin F, Reinhardt G, Gärtner S, Patyk A, Baur F, Bemmann U, Groß B, Heib M, Ziegler C, Flake M,

Schmehl M, Simon S (2004) Stoffstromanalyse zur nachhaltigen energetischen Nutzung von Biomasse. Öko-Institut Darmstadt: Endbericht.

6. Nitsch H, Osterburg B (2010) Grünland - Bestandsaufnahme und Handlungsoptionen: Erkenntnisse aus dem InVeKoS-Projekt. Experten Workshop Einsatz von Grünlandaufwuchs und Landschaftspflegematerial in. Jena: Biogasanlagen.
7. EEA (European Environment Agency) (2007) Estimating the environmentally compatible bioenergy potential from agriculture, European Environment Agency. Technical report No 12/2007. Available at http://www.eea.europa.eu/publications/technical_report_2007_12 Accessed 12 Dec 2012
8. Kopfmüller J, Brandl V, Jörissen J, Paetau M, Banse G, Coenen R, Grunwald A (2001) Nachhaltige Entwicklung integrativ betrachtet. Regeln, Indikatoren, Berlin: Konstitutive Elemente.
9. Kopfmüller J (2006) Ein Konzept auf dem Prüfstand. Das integrative Nachhaltigkeitskonzept in der Forschungspraxis. Berlin: Edition Sigma.
10. Alexandratos N, Bruinsma J (2012) World agriculture towards 2030/2050, ESA working paper no. 12–03. Agricultural Development Economics Division of the Food and Agriculture Organization of the United Nations. Available at http://www.fao.org/docrep/016/ap106e/ap106e.pdf
11. DBV (Deutscher Bauernverband) (2012) Situationsbericht 2012/13. Trends und Fakten zur Landwirtschaft. Available at http://www.bauernverband.de/situationsbericht-2013 . Accessed 12 Dec 2012
12. Osterburg B, Laggner B, Nitsch H, Röder N, Roggendorf W (2010) Analysis of grassland conversion into arable land in Northwest Germany. Available at http://www.egf2011.at/files/pubs/350_osterburg.pdf . Accessed 12 Dec 2012
13. Osterburg B, Isermeyer F, Lassen B, Röder N (2010) Impact of economic and political drivers on grassland use in the EU. Grassland Sci Europe 15:14-28
14. LEL (Infodienst Landwirtschaft-Ernährung-Ländlicher Raum, Schwäbisch Gmünd) (2012) Agrarmärkte 2011/12 Milch. Available at http://landwirtschaft-bw.de/pb/MLR.Landwirtschaft,Lde/

Startseite/Marktinformationen/Milch+_+Milchprodukte. Accessed 12 Dec 2012

15. The Federal Government (2008) Progress Report 2008 on the National Strategy for Sustainable Development. For a Sustainable Germany. Available at http://www.bundesregierung.de/Content/EN/_Anlagen/2009-05-13-fortschrittsbericht-englisch-nachhaltigkeit.pdf?__blob=publicationFile
16. Bemmann A, Feger KH, Gerold D, Grosse W, Hartmann KU, Petzold R, Röhle H, Schweinle J, Steinke C (2007) Kurzumtriebsplantagen auf landwirtschaftlichen Flächen in der Region Grossenhain im Freistaat Sachsen. Forstarchiv 78(2007):95-101
17. Wagner P, Heinrich J, Kröber M, Schweinle J, Grosse W (2009) Ökonomische Bewertung von Kurzumtriebsplantagen und Einordnung der Holzerzeugung in die Anbaustruktur landwirtschaftlicher Unternehmen. In: Reeg T, Bemmann A, Konold W, Murach D, Spiecker H (eds) Anbau und Nutzung von Bäumen auf landwirtschaftlichen Flächen, Wiley-VCH Verlag: Weinheim. pp 135-145
18. Musshoff O (2012) Growing short rotation coppice on agricultural land in Germany: a real options approach. Biomass Bioenergy 41:73-85
19. Stoate C, Baldi A, Beja P, Boatman ND, Herzon I, von Doorn A, de Snoo GR, Rakosy L, Ramwell C (2009) Ecological impacts of early 21st century agricultural change in Europe - a review. J Environ Manage 91(2009):22-46
20. Kleijn D, Kohler F, Baldi A, Batary P, Conception ED, Clough Y, DM Gabriel D, Holzschuh A, Knop E, Kovacs A, Marshall EJP, Tscharntke T, Verhulst J (2009) On the relationship between farmland biodiversity and land-use intensity in Europe. Proceedings of the Royal Society 276:903-990
21. Robert T, Watson R, Noble I, Bert Bolin NH, Ravindranath , David J, Verardo , Dokken David J (2000) Land use, land-use change and forestry. Cambridge: Cambridge University Press. p 375
22. Freibauer A, Drösler M, Gensior A, Schulze ED (2009) Das potenzial von wäldern und mooren für den klimaschutz in deutschland und auf globaler ebene. Natur und Landschaft 84(1):20-25

23. Lal R (2008) Carbon sequestration. Philos Trans R Soc B 363(2008):815-830
24. Guo LB, Gifford RM (2002) Soil carbon stocks and land use change: a meta analysis. Glob Chang Biol 8:345-360
25. Schlesinger WH (1997) Biogeochemistry: an analysis of global change. San Diego: Academic.
26. Peters W, Schultze C (2010) Bioenergie und Naturschutz. Synergien fördern, Risiken vermeiden. Bundesamt für Naturschutz (BfN): Bonn. Available athttp://www.bfn.de/fileadmin/MDB/documents/themen/erneuerbareenergien/bfn_position_bioenergie_naturschutz.pdf
27. BUND (Bund für Umwelt und Naturschutz Deutschland) (2010) Kurzumtriebsplantagen für die Energieholzgewinnung - Chancen und Risiken. Positionspapier.Available athttp://www.bund.net/fileadmin/bundnet/publikationen/landwirtschaft/20100714_landwirtschaft_bund_position_55_KUP.pdf
28. NABU (Naturschutzbund Deutschland) (2008) Energieholzproduktion in der Landwirtschaft. Chancen und Risiken aus Sicht des Natur- und Umweltschutz.Available at http://www.nabu.de/imperia/md/content/nabude/energie/biomasse/nabu-studie_energieholz.pdf
29. SRU (Sachverständigenrat für Umweltfragen) (2007) Klimaschutz durch Biomasse. Stuttgart: Sondergutachten.Available athttp://www.umweltrat.de/cae/servlet/contentblob/467474/publicationFile/34339/2007_SG_Biomasse_Buch.pdf
30. Ledin S (1998) Environmental consequences when growing short rotation forests in Sweden. Biomass Bioenergy 15:49-55
31. Ranney JW, Mann LK (1994) Environmental considerations in energy crop production. Biomass Bioenergy 6:211-228
32. Sage RB, Cunningham M, Boatman N (2006) Birds in willow short-rotation coppice compared to other arable crops in central England and a review of bird census data from energy crops in the UK. Ibis. 148:184-197
33. Rossi A (2012) Good environmental practices in bioenergy feedstock production. Making bioenergy work for climate and food security. Environment and Natural Resources management Working Papers 49.Available at http://www.fao.org/docrep/015/i2596e/i2596e00.pdf

34. Pellegrino E, Di Bene C, Tozzini C, Bonari E (2011) Impact on soil quality of a 10-year-old short-rotation coppice poplar stand compared with intensive agricultural and uncultivated systems in a Mediterranean area. Agric Ecosyst Environ 140(1–2):245-254
35. Fry DA, Slater FM (2011) Early rotation short rotation willow coppice as a winter food resource for birds. Biomass Bioenergy 35(7):2545-2553
36. Goransson G (1994) Bird fauna of cultivated energy shrub forests at different heights. Biomass Bioenergy 6:49-52
37. Berg A (2002) Breeding birds in short-rotation coppices on farmland in central Sweden-the importance of Salix height and adjacent habitats. Agri Ecosyst and Env 90:265-276
38. Sage RB, Tucker K (1997) Invertebrates in the canopy of willow and poplar short rotation coppices. Aspects Appl Biol 49:105-111
39. Sage RB (1998) Short rotation coppice for energy: towards ecological guidelines. Biomass Bioenergy 15(1):39-47 t
40. Lettens S, Muys B, Ceulemans R, Moons E, Garcia J, Coppin P (2003) Energy budget and greenhouse gas balance evaluation of sustainable coppice systems for electricity production. Biomass Bioenergy 24(3):179-197
41. Matthews R, Robinson R, Abbott A, Fearis N (1994) Modelling of carbon and energy budgets of wood fuel coppice systems. London, UK: Department of Trade and Industry.
42. Dubuisson X, Sintzoff I (1998) Energy and CO_2 balances in different power generation routes using wood fuel from short rotation coppice. Biomass Bioenergy 15(4–5):379-390
43. DBFZ (Deutsches BiomasseForschungsZentrum (2011) Monitoring zur Wirkung des ErneuerbareEnergien-Gesetz (EEG) auf die Entwicklung der Stromerzeugung aus Biomasse.Available athttp://www.dbfz.de/web/fileadmin/user_upload/Userupload_Neu/Stromerzeugung_aus_Biomasse_Zwischenbericht_Maerz_2011.pdf
44. Bell S, McIntosh E (2001) Short rotation coppice in the landscape. UK.Available at http://www.forestry.gov.uk/PDF/fcgn2.pdf/$FILE/fcgn2.pdf

45. Petzold R, Feger KH, Siemer B (16/2006) Standörtliche Potenziale für den Anbau schnellwachsender Baumarten auf Ackerflächen. Allg. Forst. Z. Waldwirtsch. Umweltvorsorge 16/2006:855-857
46. Murach D, Hartmann H, Walotek P (2008) Ertragsmodelle für landwirtschaftliche Dendromasse. Endbericht DENDROM - Zukunftsrohstoff Dendromasse. Berlin, Cottbus: Ebers-walde. pp 146-149
47. Aust C (2012) Abschätzung der nationalen und regionalen Biomassepotentiale von Kurzumtriebsplantagen auf landwirtschaftlichen Flächen in Deutschland, Dissertation an der Fakultät für. Forst- und Umweltwissenschaften: Universität Freiburg.
48. Seitz R (2010) Rinderhaltung in Baden-Wuerttemberg. Statistisches Monatsheft Baden-Wuerttemberg. 30-35
49. Rösch C, Raab K, Stelzer V (2005) Surplus grassland - a new source of bio-energy? In: Proceedings of the 14th European biomass conference & exhibition. Paris: Frankreich. pp 312-315 CD-ROM
50. Landwirtschafts- und Landeskulturgesetz Baden-Württemberg Novellierung vom 17.12.2011. Available at http://www.landesrecht-bw.de/jportal/?quelle=jlink&query=Lw%2FKultG+BW&psml=bsbawueprod.psml&max=true&aiz=true*webcite*
51. Ministerrat der Europäischen Union (1992) Richtlinie/92/43/EWG vom 21. Mai 1992 zur Erhaltung der natürlichen Lebensräume sowie der wildlebenden Tiere und Pflanzen. Available at http://eur-lex.europa.eu/LexUriServ/LexUriServ.do?uri=CONSLEG:1992L0043:20070101:DE:PDF
52. Oppermann R (2009) Grünland und Schutz von Biodiversität und Gewässern. Vortrag auf der Tagung 'Naturschutz und Landwirtschaft im Dialog - Grünland im Umbruch' am 27.-30.04.2009 an der Internationalen Naturschutzakademie Insel Vilm des Bundesamtes für
53. Richtlinie des Ministeriums Ländlicher Raum zur Förderung der Erhaltung und Pflege der Kulturlandschaft und von Erzeugungspraktiken, die der Marktentlastung dienen (Marktentlastungs- und Kulturlandschaftsausgleich – MEKA II –) vom 12.09.2000. Available at http://www.vd-bw.de/spiderlink/7803-3-1-3alt0708_02.B_inhalt

54. Tonn B, Messner J, Elsässer M (2010) Einsatz von (MEKA-) Grünlandaufwüchsen in Biogasanlagen in Baden-Württemberg, Talk on Expert Workshop. Jena: Einsatz von Grünlandaufwuchs und Landschaftspflegematerial in Biogasanlagen. Available at http://www.tll.de/ainfo/pdf/gland/gn10_06f.pdf
55. Hartmann A (2008) Wie viel Fläche wird für Biogas benötigt? Statistisches Monatsheft Baden-Württemberg 7/2008. 40-42
56. Gömann H, Kreins P, Münch J, Delzeit R (2010) Auswirkungen der Novellierung des Erneuerbare-Energien-Gesetzes auf die Landwirtschaft in Deutschland, talk on the GEWISOLA meeting. Möglichkeiten und Grenzen der wissenschaftlichen Politikanalyse, 29.09-01.10. Braunschweig
57. Breustedt G, Habermann H (2010) Einfluss der Biogaserzeugung auf Landwirtschaftliche Pachtpreise. talk on the GEWISOLA meeting. Möglichkeiten und Grenzen der wissenschaftlichen Politikanalyse. Braunschweig
58. Statistisches Landesamt Baden-Württemberg (2010) Milcherzeugung und Milchverwendung in Baden-Württemberg. Available at http://www.statistik.baden-wuerttemberg.de/veroeffentl/Statistische_Berichte/3432_10001.pdf

Chapter 8

Assessing Stakeholders' Perceptions and Values Towards Social-Ecological Systems Using Participatory Methods

Grace B Villamor[1,2], Ignacio Palomo[3,4], Cesar A López Santiago[4], Elisa Oteros-Rozas[4], and Joe Hill[5]

[1]Department of Ecology and Natural Resources Management, Center for Development Research (ZEF), University of Bonn, Bonn, Germany

[2]World Agroforestry Centre (ICRAF), Jl. Situ Gede, Bogor, 16115, Indonesia

[3]Basque Centre for Climate Change (BC3), Alameda Urquijo 4, Bilbao, 48008, Spain

[4]Department of Ecology, Social-Ecological Systems Laboratory, Universidad Autónoma de Madrid, Madrid 28049, Spain

[5]Department of Political and Cultural Change, Center for Development Research (ZEF), University of Bonn, Bonn, 53113, Germany

ABSTRACT

Introduction

Many conservation initiatives for managing social-ecological landscapes factor in the best available biophysical information. However, insufficient consideration of social aspects can render such initiatives ineffective. By incorporating stakeholders' perceptions and values, and by involving stakeholders (transparently) in decision-making processes, conservation plans and efforts can better achieve desired goals and targets.

Results

In this paper, we present and analyze three case studies within different ecological and land use contexts. Each case study uses different participatory approaches to elicit stakeholders' perceptions and values, while seeking to involve stakeholders in the decision-making process. The case studies are the following: (1) participatory scenario planning for a protected area in Doñana coastal wetland in Southwestern Spain; (2) assessment of ecosystem services trade-offs and social responses on rewards for agro-biodiversity in the rubber agroforest landscape in Jambi, Indonesia; and (3) socio-cultural evaluation, through visual stimuli, of ecosystem services provided by transhumance cultural landscapes in Central Spain. We discuss how stakeholders perceive and value their environments and to what extent participatory approaches are useful for capturing information relating to land use and ecological processes.

Conclusions

Social-ecological systems are inherently complex, having a variety of interacting actors with different types of ecological knowledge, interests, and values. Different participatory tools or approaches are appropriate for various and specific contexts and objectives. Determining and integrating the various types of knowledge and values of different actors can contribute to a more comprehensive understanding of social-ecological systems.

INTRODUCTION

Social-ecological system (SES) frameworks are being promoted for understanding ecological processes at the landscape level, in recognition that insufficient consideration of social systems can render the management or conservation of such ineffective systems (Mascia et al. [2003]; Ban et al.[2013]). SES frameworks conceptualize human/social and natural/ecological systems as coupled systems in which people interact with natural components, drawing on theories about the co-evolutionary nature of human and biophysical systems (Norgaard [1994]). The study of the complex interactions between the biophysical and social components of SES, conceptualized as a set of subsystems, can provide a deeper knowledge of the overall system. SES frameworks' major features include the analysis of complex interactions and feedbacks between human and natural systems, the integration of tools and techniques from both the ecological and social sciences, and require an interdisciplinary study team to address common research questions (Liu et al. [2007]). Moreover, the frameworks entail the study of ecosystem services through socio-cultural assessments, crucial to identifying the perceptions of individuals, institutions, and organizations towards ecosystem services, based on their preferences, needs, values, norms, and aspirations (Cowling et al. [2008]).

In confronting the challenges to the planning and management of SESs, we need to deal with a variety of actors or stakeholders such as scientists, policy makers, private sector actors, and community-based resource users. By incorporating the stakeholders' perceptions and values, and by seeking to involve stakeholders (transparently) in decision-making processes, conservation plans and efforts will likely better achieve desired goals and targets (Young et al. [2013]). Scientific information that we feed into the planning and assessing exercises is likely to be effective in influencing the public only if that information is perceived by the relevant stakeholders as legitimate, credible, and salient (Cash et al. [2003]). In most cases, disputes exist among different stakeholders due to differences in their interests, values, and goals (i.e., short-term vs. long-term goals or recreational over subsistence needs).

Thus, careful consideration of differing viewpoints is necessary to avoid biased decisions or top-down policy-led or science-led judgments regarding the management of SES.

To date, most ecosystem service studies have focused primarily on monetary and biophysical approaches to valuation/assessment, while few studies have explored the more intangible socio-cultural perceptions and preferences (Nieto-Romero et al. [2013]; Vihervaara et al. [2010]). Valuation methods are not ideologically neutral (Gómez-Baggethun et al. [2010]), but rather culturally (and politically) constructed and, as such, act as value-articulating institutions that influence the articulation of decision-making processes related to the environment (Vatn [2005]). Nevertheless, considering values is a way to understand what matters to people and the motivations underlying their decisions. Moreover, a consideration of values allows for the differentiation of ecosystem services, benefits, and values that are often conflated in the literature (Chan et al., [2012]). However, it is very difficult to encompass the multiple dimensions of values systems in a single methodological approach. Any value-eliciting approach may tackle a complex combination of various dimensions of values, which are always context dependent. Chan et al. ([2012]) present a comprehensive typology of value dimensions, connected to a set of ecosystem services and benefits, and advocate a multi-method approach to better match the diversity of values linked to the interplay of ecosystems and societies.

SES thinking has remained superficial in many studies. For example, some of the tools or approaches to investigate the interactions in SESs tend to promote opposing views of nature and society. The properties of ecosystems and societies are supposed to be considered equally, and recognition is given to normative factors (e.g., cultural values, power relations) for understanding social change (*ibid.*). Moreover, (physical) scientists have perceived the study of ecological processes using quantitative and top-down methods to be precise and the analysis of social processes by qualitative, bottom-up methods to be less precise. More recently, it has been recognized and accepted that the use of qualitative methods (including participatory approaches) is valid and necessary for the evaluation of ecological processes and values. A growing number of qualitative and semi-quantitative methods are now in use; however, certain issues still need to be resolved. For example, the metrics of preference in surveys of individual choice, and the use of subjective scaling to understand participative and deliberative processes, need to be compared to acknowledge their complementarities as well as the appropriate conditions under which these methods are most efficient. For better environmental management, policy makers and decision makers need a variety of tools and methods that allow social and ecological analyses of social-ecological processes and systems.

In this paper, we present and analyze three case studies within different ecological contexts and land uses, each of which use participatory approaches to draw out stakeholders' preferences and perceptions and to involve stakeholders in the decision-making process. The case studies are the following: (1) participatory scenario planning for a protected area in Doñana coastal wetland in Southwestern Spain; (2) assessment of ecosystem services trade-offs in a rubber agroforest landscape in Jambi, Indonesia; and (3) assessment of ecosystem services using a visual-stimulus approach in the transhumance cultural landscapes of Central Spain. Our main objective is to compare and contrast several participatory approaches used in environmental assessments in order to assess their main advantages and limitations. The specific objectives of this study are as follows: (1) to describe the application, in different contexts, of several participatory methods, namely participatory scenario planning, participatory ecosystem service assessment of trade-offs, and participatory evaluation of ecosystem services through visual preferences; (2) to evaluate the extent to which ecological aspects as articulated by stakeholders were assessed by each method; and (3) to assess the different value conceptions captured by each methodology and how this might influence overall results.

METHODS

To achieve the above-mentioned objectives, we have prepared a conceptual framework (Figure 1) that captures the degree of participation that any method allows, models in a simplistic three-fold manner the stakeholders involved, according to the types of knowledge they possess, and provides a schematic detailing of the various dimensions of values that stakeholders can possess. In this section, we elaborate upon these three components, while in the subsequent section we describe in depth each of the three case studies, the participatory methods implemented and the key actors involved, with summaries of the results. Then follows a discussion and the conclusion.

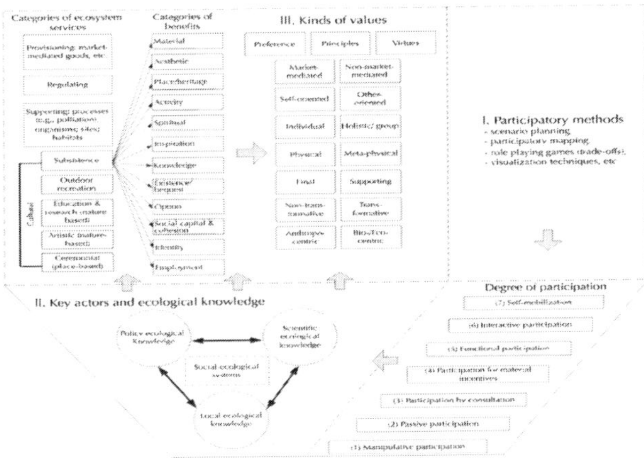

Figure 1: Conceptual framework. Conceptual framework integrating (I) degrees of participation the methods provide Pretty et al. ([1995]:61), (II) types of ecological knowledge of different actors, and (III) eight dimensions of values related to different categories of ecosystem services and benefits (Chan et al. [2012], p.13).

To analyze the degree of participation a method allows, we follow the typology described by Pretty et al. ([1995]:61), elaborated as follows:

Manipulative participation - participation is a pretense, stakeholders have unelected representatives;

1. *Passive participation* - stakeholders are told what has been decided or happened while information being shared belongs only to professionals (e.g., scientists/researchers, extension services personnel);
2. *Participation by consultation* - stakeholders are consulted or asked to answer questions. External actors define the problems and information gathering processes, and control the analysis;
3. *Participation for material incentives* - actors contribute resources (e.g., labor) in return for food, cash, or other materials, but are not involved in either the experimentation or the process of learning;
4. *Functional participation* - actors are involved in forming groups to meet predetermined objectives related to a project or experiment;

5. *Interactive participation* - actors participate in joint analysis, seek multiple perspectives, and make use of systematic and structured learning processes; and
6. *Self-mobilization* - actors take initiatives independently of external institutions to change systems, develop contacts with other external actors for technical advice and resources, while retaining control over how resources are used.

To conceptually separate the stakeholders and their knowledge systems, we use a three-fold distinction of ecological knowledge, distinguishing between local, policy, and scientific. This three-fold schematic was developed by the lead author in her Indonesian research, and perhaps fits less well to the two Spanish case studies due to the different social and political economic conditions between these countries. The three knowledge types are the following:

1. *Local ecological knowledge* - the rich cumulative body of knowledge, practice, and belief evolving by adaptive processes and handed down through generations by cultural transmission, about the relationship of living beings with one another and with their environment (Berkes and Folke [1998]; Berkes et al. [2000]). Although many studies have explored the role of local ecological knowledge in perceiving and understanding diverse environmental features and changes, fewer studies have actually linked these cognitions with measured changes in behaviors (Berkes et al. [2000]).
2. *Policy ecological knowledge* - ecological knowledge shaped by the rules, regulations, and maps that exist for politically defined territorial and natural resources jurisdictions. The actors having such knowledge include government agencies, policy makers, local authorities, and non-government organizations (NGOs), mostly engaged in projects with short-term impacts. However, they may have access to data that is otherwise unavailable to local people (Voinov and Gaddis [2008]). This data can often be provided to scientists or researchers if permission is granted.
3. *Scientific ecological knowledge*- a type of knowledge generated from the representation of complex systems and processes and analyzed through scientific approaches, e.g., deductively and/or inductively to seek generic mechanisms (van Noordwijk [2011]).

We adopt Chan et al.'s ([2012]) eight dimensions of values to analyze the values elicited during the decision-making processes. Of the existing literature on ecosystem services values, we consider Chan et al. ([2012]) to be the most comprehensive particularly on conforming to cultural services (as well as non-use values). Figure 1 shows that since many ecosystem services co-produce cultural benefits, full characterization of services (especially non-material values) is only possible using diverse methods drawn from the social sciences.

The eight dimensions of values include the following:

1. *Preferences vs. principles vs. virtues* - follows a division of ethical theories;
2. *Market-mediated vs. non-market-mediated* - this differs from the market/non-market valuation dichotomy of economics;
3. *Self-oriented vs. others-oriented* - distinguishes concern for oneself vs. others, thus for example, allowing perspectives of those who cannot represent themselves to be taken into consideration;
4. *Individual vs. holistic/group* - most valuation methods orient towards one or the other;
5. *Experiential/physical vs. metaphysical* - objects valued for their contribution to valued existence and for their existence independent of experience;
6. *Supporting vs. final (instrumental vs. inherent)* - things valued because they help to produce things, other values inherent in that they are the desired ends;
7. *Transformative vs. non-transformative* - thing or process valued for its contribution to a transformation in values, or valuable in reference to unchanging values;
8. *Anthropocentric vs. biocentric/ecocentric* - values held by humans or non-humans.

RESULTS

(1) Participatory Scenario Planning for a Protected Area in Doñana Social-Ecological System in Southwestern Spain

The Doñana social-ecological system (Doñana SES) consists of four ecodistricts (marsh, aeolian sheets, estuary, and coast) (Montes et al., 1998). The Doñana marsh, covering 1,660 km^2, is one of Europe's biggest coastal wetlands and maintains a high biodiversity that delivers multiple ecosystem services, including biodiversity conservation, nature tourism, and water regulation (Zorrilla-Miras et al. [2014]). The social system includes several institutions, some of which have shaped the past and present of Doñana, such as the Doñana Protected Area. Nonetheless, in the previous decades, much of the marshes were transformed into agricultural lands. As a result, the protected area is surrounded by a matrix dedicated to agriculture which mainly delivers food (Palomo et al. [2014]). The most important crops being produced are rice and red fruits, many of which are exported to other countries, while non-irrigated farming has declined in the last decades.

Although the earlier abrupt land use changes have now stabilized, many uncertainties still exist such as the effects of climate change, recent mine spills from a private company operating nearby and excessive and unexplained mortalities of fauna in the recent past. Moreover, there are several conflicting interests in the region, and there is a need to create shared visions of the future for the region. These were the reasons that led to a participatory scenario planning process for assessing uncertainty in the future evolution of Doñana and to propose pathways towards a desirable future until 2035 (Palomo et al. [2011]). The project was funded by the World Wildlife Fund (WWF), one of the most active NGOs in the region.

i. Participatory Research Methods

The methodology employed in the participatory scenario planning is described in Palomo et al. ([2011]). Participants worked in groups in which professional facilitators helped guide the process. Each group comprised actors from all three knowledge categories (Figure 1), in order to foster discussion between different interest groups and to reach consensus. The two workshops lasted one and a half days in total and included three phases: from the past to the present, scenario building, and back-casting. The past to the present phase included a participatory assessment of the evolution of Doñana SES during the last decades. This was followed by the scenario building exercise, in which stakeholders created four scenarios for the Doñana SES towards the year 2035. Finally, the back-casting methodology was applied in order to obtain from participants management recommendations that would facilitate approaching a desirable future, taking into account the four scenarios created. In the future scenarios, besides social or economic aspects, the actors were asked to describe different ecological characteristics such as the state of the wetland, biodiversity, or land uses across the landscape.

ii. Actors and Preferences

A total of 52 actors participated in the two workshops in 2009. The diverse background, values, and preferences of the participants influenced the workshops' outcomes. In general, the preferences of stakeholders ranged between those who favor economic development driven by agriculture and tourism and those supporting a better conservation of the area. Many showed intermediate preferences, located between these extreme visions. The actors can be grouped in the following categories, although it can be noted that in many cases these categories overlap:

Actors with local ecological knowledge: farmers (7), livestock raisers (4), hunters (2), and honeybee keepers (1). The preference of this group was mainly towards regulations that allowed the continuance of their livelihood practices and hence, less legal restrictions. Some stakeholders, such as the organic farmers, were concerned with environmental problems and supported more strict ecological production standards. Other actors invited to the workshops, such as professional shellfish gatherers, did not attend.

Actors with policy ecological knowledge: managers from the protected area (7), administrators from the municipalities (7), regional water agency (2), regional environmental office (3), local sustainability agency (3), NGO representatives (3), environmental education professionals (3), representatives of the private sector related to tourism (2), and journalists (2). In this group, there was a wide range of preferences, among which many actors had a preference for an 'intermediate' outcome between conservation and development.

Actors with scientific knowledge: researchers from different universities and research centers (6). Those researchers who focused on biological aspects had preferences that tended to more environmentally sustainable outcomes. The social scientists' preferences were also oriented to sustainable results but in a more socially inclusive manner. In this sense, one aspiration of the researchers was to reach a consensus among participants for what they perceived to be a sustainable future.

iii. Ecological Aspects Assessed

Different ecological aspects, mainly related to biodiversity, water, agriculture, and tourism, were discussed and analyzed during the workshops. Moreover, since the state of ecosystem services was assessed in each scenario, underlying ecological aspects were discussed by the participants regarding their interconnections with social aspects and human well-being. For example, when issues dealing with water quality and biodiversity inside the protected area were discussed, these were related to other aspects such as use of water in agriculture, water management in the watershed, or the transportation sector in the area which is responsible for high mortality of fauna.

iv. Main Values Elicited

The kinds of values elicited within a deliberative process are very diverse. Given the nature of the method used, in which ecosystem services trends were assessed and linked to human well-being, there was a broad space for the inclusion of multiple values (though note that monetary valuation of ecosystem services was not performed). As most of the outcomes of the workshops were obtained deliberatively, group values as well as others-oriented values (as opposed to individual values) were emphasized. The medium- to long-term perspective that the scenario method brings into the discussion most probably facilitates the inclusion of supporting and final values as the focus is in a story instead of in a snapshot (which would be favoring only the final values).

(2) Participatory Assessment of Ecosystem Services Trade-offs for a Rubber Agroforest Landscape in Jambi, Indonesia: Mapping and Role-Playing Game

In Jambi province, Sumatra, Indonesia, rubber agroforests are being replaced by export-oriented monoculture plantations such as rubber and oil palm. The rubber agroforest, or jungle rubber system, is a complex multi-strata canopy that is similar to natural secondary forests (Gouyon et al. [1993]; Michon [2005]). It also supports lowland tropical forest diversity, as it serves as a corridor for diverse flora and fauna and represents a substantial carbon stock. Since most of the lowland tropical forests in Sumatra have been converted to plantations of rubber and oil palm, assessing the possible ecosystem services trade-offs during the conversion of rubber agroforests is crucial to support decision-making. The study area is located in the foothills of Kerinci-Seblat Natural Park in Bungo district, Jambi Province. Three villages (Desa Buat, Lamang Panjang and Lubuk Beringin), containing the last remaining rubber agroforests in the province, were investigated.

i. Participatory Research Methods

Participatory mapping and land use role-playing games at the watershed level (total area of 160 km^2), were used to determine the local perceptions of rubber agroforest farmers. The results of these approaches, combined with a household survey (N=196), were incorporated in an agent-based model (Villamor [2012]) to assess the potential trade-offs of introducing payment for ecosystem services (PES) schemes.

The aim of the participatory mapping was to integrate local stakeholders' perceptions of the different land use types and institutional properties (i.e., ownership, customary, protected, and production areas). Google Earth maps were used, which give good visual images of the actual land cover in the area, with key landmarks (e.g., roads, bridges, schools, and houses) clearly visible. A total of 95 household respondents, including the village heads and a representative of Warung Konservasi (WARSI, a local NGO) jointly identified household farm plots. Placing tracing paper over the map, respondents were able to sketch the roads to their farm plots, their neighboring farms, and other relevant spatial information (i.e., distance of market or pick-up stations). Later, the data were processed using GIS and exported to Netlogo (i.e., an agent-based model platform) (Villamor et al. [2014]).

The aim of the role-playing game was two-fold: 1) to identify the preferred land-use types that meet villagers' objectives under varying scenarios, and 2) to determine the social response of villagers to external actors promoting conservation of rubber agroforests through payments for ecosystem services schemes or conversion of rubber agroforest into oil palm and rubber plantations (Villamor and van Noordwijk [2011]; Villamor et al. [2013]). In the game, three land-use game boards, which represented each village landscape, were used. The players were the same respondents who joined in the participatory mapping exercise. Using the game boards as simulators, the players could directly decide and change the village landscape in response to their negotiations with external agents. Roles such as oil palm and logging concessionaires, government, local NGO (promoting biodiversity conservation) representatives were played by the villagers, matching as much as possible the real life roles of external agents. To make the game more realistic and to test how land-use preferences change while meeting people's targets, we integrated some scenarios such as socio-economic shocks and natural calamities. During the game, only village leaders were involved; no private sector actors were invited to participate.

ii. Actors and Preferences

The following actors were involved:

Actors with local ecological knowledge are the men and women in the villages. The majority of the men are rubber tappers while women are responsible for the production of the staple food, rice. Though the area is endowed with rich natural resources, these villages are considered poor, having little access to markets, roads, communication, and electricity infrastructure. Reciprocity and trust are high in the area

(Akiefnawati et al. [2010]). Research has revealed that men prefer to conserve both forests and rubber agroforests as a source of income and timber. Women prefer to maintain rubber agroforests, which supports the supply of the water needed to generate electricity from mini-hydroelectric power, to irrigate rice fields and to supply drinking water (Villamor et al. [2013]). However, due to low yields and low financial returns from rubber agroforests as compared to monoculture rubber and oil palm, the villagers are attracted to convert their agroforests to monoculture.

The actor having policy ecological knowledge is the provincial government of Jambi that aims for economic development of its territorial domain. Officially, it considers the monoculture rubber and oil palm plantations as engines for economic growth in the province. Nevertheless, environmental conservation, via support of protected areas as the easiest and cheapest option, is also its stated objective.

Actors with scientific ecological knowledge are the conservation agencies such as the WARSI, the World Agroforestry Centre, and the lead author of this paper (referred to as researcher in the remaining part of this section), who aim to reconcile the land use preferences of the local people and the provincial government. These actors prefer forests and agroforests due to the rich biodiversity such land uses support, and perceive the establishment of a PES scheme as a win-win option especially for rubber agroforests.

iii. Ecological Aspects Assessed

Soil restoration (e.g., fallow practices), flood regulation as maintained by trees, biodiversity functions (e.g., pest control and predation), and regeneration of high-valued timber trees are the ecological functions and processes most often expressed by actors during the role-playing games and participatory mapping activities. Women are mostly interested in maintaining a certain distance between their village vegetable gardens and rubber agroforests because wild boars may destroy their vegetables. On the other hand, the majority of men prefer to maintain blocks of forest and agroforest within their landscapes because of their use of timber for house construction. Both males and females tend to locate environmentally destructive land use such as coal mining away from their village settlements. It was also observed that actors at the local level have a strong explicit understanding of ecological processes as well as spatial relations (e.g., upstream and downstream water flows).

Researcher and respondents agreed that microclimate, water regulations and water quality are among examples of indirect benefits from rubber agroforests. These correspond with the findings of scientific studies that show how rubber agroforests' functions resemble those of secondary forests' (Long and Nair [1999]; Beukema et al. [2007]; Gouyon et al. [1993]; Tomich et al. [2001]). Furthermore, the role-playing games highlighted the differences in land-use preferences between females and males - that females preferred more profitable land uses (e.g., oil palm) while males preferred forest and rubber agroforests (Villamor et al. [2013]).

iv. Main Values Elicited

Although the majority of the values elicited fall into the material aspects (i.e., goods such as timber, food, and water) as anthropocentric value (Figure 1), a transformative kind of value (i.e., if it is valuable in reference to unchanging values and perspectives, Chan et al. ([2012])) was drawn out from the mapping and role-playing games. For example, the wild boar might be seen as a pest from the women's perspective, while overall there is no preexisting preference for or against preservation of wild boar, since the species is not part of the local peoples' dietary intake -due to religious reasons- and thus may be perceived as valueless. On the other hand, the wild boar is part of the food web of the larger animals (e.g., Sumatran tigers); therefore, its presence maintains a balance in the biological system and so the value of the wild boar is bio-centric (Figure 1). Another interesting value elicited from the game is that of non-market-mediated. The study villages have a matrilineal inheritance system for their rice paddies, and (one of the ways) men claim land rights (is) by planting rubber trees (van Noordwijk et al. [2012]), thus as compared to women, men showed a preference for more rubber agroforest.

(3) Assessment of Ecosystem Services Using a Visual-Stimulus Approach in the Transhumance Cultural Landscapes of the Central Spain (Cuenca and Ciudad Real Provinces)

Transhumance is a customary pastoralist practice consisting of the seasonal migration of livestock between ecological regions following peaks in pasture productivity (Ruiz and Ruiz [1986]). A network of legally protected drove roads connects summer and winter pastures, which are the most outstanding feature of transhumance landscapes. This livestock movement has declined in Spain since the eighteenth century (Ruiz and Ruiz [1986]; García-Martín [2004]; Oteros-Rozas et al.[2013]), but some herds still continue to migrate on foot. This is the case of the Conquense Drove Road (CDR), one of the longest and most frequently walked drove roads today. CDR consists of a 75-m wide and approximately 410-km long corridor that crosses a mosaic of agroecosystems comprising cereals, sunflowers, vineyards, olive groves, some remnants of Mediterranean oak forest in Castilla-La Mancha (Cuenca and Ciudad Real provinces), and pine forests in the Serranía de Cuenca and Sierra de Albarracín (Cuenca and Terual provinces). The study area was characterized as a social-ecological network (Oteros-Rozas et al. [2012b]).

Transhumance has been acknowledged as an important traditional farming practice for provisioning services such as wool and high-quality meat, regulating services such as seed dispersal, and cultural services such as cultural identity and traditional ecological knowledge, while also contributing to biodiversity conservation (Gómez Sal and Lorente [2004]; Bunce et al. [2006]; Manzano and Malo[2006]; Oteros-Rozas et al. [2012a]). Experts consider the livestock route as a key corridor linked to landscape connectivity playing an important role on cultural landscape sustainability and biodiversity conservation.

Using a visual-stimulus approach (photographs), we explored the preferences and perceptions of stakeholders about the ecosystem services provided by two different landscapes and the perceived effect of the Conquense Drove Road (López-Santiago et al. [2014]). The project within which this research was embedded had the main objective of identifying and assessing ecosystem services provided by transhumance and related ecosystems, to provide insights for decision-making that allow the maintenance of drove roads and transhumance. The precise aim of this research was to elicit actors' preferences and perceptions of ecosystem services through use of visual stimuli.

i. Participatory Research Methods

The visual method used has been recognized as an effective research tool for the study of human-ecosystem relationships (e.g., Daniel [2001]). Based on landscape perception surveys, the visual-stimuli approach was employed to explore the preferences and perceptions of ecosystem services provided by the two main landscapes (croplands and forests) crossed and connected by the CDR (López-Santiago et al. [2014]). Following Pretty et al's ([1995]) typology, participation is in this case *'by consultation'* and designed to inform decision makers about the preferences, perceptions, and values of citizens and scientists strongly linked to the research problem.

Data were obtained in the form of standardized questionnaires (N=314) applied in face-to-face interviews. Participants were shown two pairs of very similar photographs (one pair for a cropland and the other for a pine forest), with each picture in a pair differing by the presence or not of the drove road. The participants were asked to state their preference for one photograph of each pair and also to score their perceived delivery of a list of 16 ecosystem services (previously identified through a literature review and in-depth semi-structured interviews in another phase of the research (Oteros-Rozas et al., [2012b])), using a semi-quantitative scale from 1 (nothing) to 4 (very high). Quantitative analysis used descriptive statistics (mean frequencies and standard deviations), and Wilcoxon's rank-sum tests were performed to identify and describe differences in ecosystem service perceptions between landscapes and the presence of the CDR. To explore how underlying socio-cultural variables might have influenced the perception of ecosystem services, we applied a multivariate model of redundancy analysis (Rao [1964]) that showed the relationship between perception values and the socio-cultural attributes of respondents.

ii. Actors and Preferences

The participants of the survey were local inhabitants, regular visitors, and scientists:

Actors with local ecological knowledge and values: herders and the people residing in the small villages in the transhumance summering area (Cuenca) showed a marked preference for landscapes with drove roads and appreciated pastures, livestock, fire prevention, connectivity, esthetic value, cultural identity, and tourism. Contrastingly, farmers living in La Mancha (Ciudad Real) (a cropland landscape crossed by the drove road) mostly appreciated their cultural identity and hunting. They did not exhibit a preference for the drove road landscape and saw more esthetic value when it is absent. Showing yet further diversity within this grouping, local people living in nearby small cities, as well as regular visitors from larger cities, preferred views of the landscape without a drove road. They related such views with the provision of tranquility/relaxation, woodland, air purification, and plant regeneration.

Actors with policy knowledge and values: village heads and staff of regional development agencies, NGOs, and government administration were approached and interviewed during other phases of the larger research project. For this particular exercise, we did not survey them because the focus was to explore perceptions among actors linked to the study area (e.g., local inhabitants and visitors) and among environmental students.

Actors with scientific knowledge and values: environmental sciences researchers and students were attracted by the presence of a drove road but were of the view that ecological connectivity, cultural identity, and esthetics are greater when a drove road is absent.

iii. Ecological Aspects Assessed

The visual pair-wise comparison methodology creates a preference index through a choice act based on paired comparison (with or without the presence in the pictures of key landscape elements). At the same time, it gives participants the opportunity to assign a value to many ecological aspects, as they were directly interrogated about their perceptions of the delivery of ecosystem services. Many of these aspects were directly linked to ecological functioning (regulating services), such as air purification, plant regeneration, soil erosion, habitat for species, connectivity, or fire prevention; and some others were indirectly linked as provisioning ecosystem services for humans: gathering, feed for animals, food from agriculture, wood and timber, and livestock.

iv. Main Values Elicited

The main values elicited from this case study for each stakeholder group (according to Chan et al.[2012], Figure 1) were the following:

Herders and locals of small villages at the study area exhibit values associated to pastoralism and linked to their traditional cultural identity -'non-transformative'- still as part of their livelihood linked to livestock, currently mediated also by tourism, outdoor recreation, educational, or other cultural demands. It is the most complex and multiple set of values found.

La Mancha's farmers have a strong cultural identity, values, and feelings linked to their local cropland-like landscape. Hunting is an important part of this. The farmers also have a complex mixture of values similar to those of the herders but with important context-dependent preferences for Mediterranean crops (vineyards and olive groves) and lesser affinity for (if not some aversion to) transhumance and livestock. Given the local economy, we could say that these are largely 'physical' values.

Environmental-friendly people show highly 'bio-centric' values and it seems as if a conflict arises between this way of thinking and their knowledge of transhumance's positive ecological impact.

Urban people's preferences were indicators of recreational-consumerist demands, mostly 'self-oriented' and 'metaphysical'.

DISCUSSION

Several authors have highlighted the need for participatory approaches to be used to improve environmental decision-making (Reed [2008]; Ban et al. [2013]). Comparison of the three case studies presented here allows us to identify some of the pros and cons of each of the participatory methods used. However, this comparative assessment is limited because the three methods were applied in different socio-ecological contexts, and therefore, differences across them are caused by both the methods and the contexts. Nonetheless, we consider that the three methods presented are different enough to allow a cross comparison, considering them as complementary.

The Doñana case study might be considered a functional participatory approach (see Reid et al.,[2009]). The main advantage of such participatory scenario planning (PSP) is that it allows an integrative analysis of complex systems, addresses uncertainty, and can be combined with a back-casting technique to develop concrete management proposals. In PSP, ecological and social considerations are assumed to be interlinked and affecting multiple aspects in the scenarios. This process is therefore useful for complex systems that face high uncertainty. An advantage of the approach is that all kinds of stakeholders might participate in the process because no formal knowledge is needed. However, given that many issues are analyzed, it is also possible that few of them are analyzed in very much detail. In addition, the PSP process is very demanding in terms of time and team effort. Thus PSP can be applied as a general approach that uncovers the main values, aspirations, fears and trade-offs between social groups, and can be complemented with other more specific methods to allow for a more in-depth focus on specific concrete issues.

Supporting/final values elicited by PSP can be considered as a powerful asset to understand complex issues and foster long term thinking. Group values might facilitate the achievement of decisions by consensus, making the method appropriate for SES in which stakeholders have *a priori* divergent interests (Vervoort et al. 2014). In relation to the three knowledge types, some of the participants that we grouped within the 'local ecological knowledge' were older than those in the other two knowledge groups and, as a result, felt less comfortable in the participatory workshops. Therefore, although this methodology is inclusive of different knowledge types, it is still important to choose adequate methodologies within PSP and to facilitate the inclusion of all knowledge types into the discourse during the participatory process.

The role-playing games were used in Indonesia's Jambi province to understand whether the land-use preferences of households and village heads aligned with those other actors. Since the concept of ecosystem services is vague for local people (and no local translation is available), the term was translated in a way that local people could understand. For this reason, the concept was explored by contrasting the direct and indirect benefits (*lansung* or *tidak langsung* in Bahasa Indonesia) of each land use (Figure 1). Nevertheless, because role-playing games are highly interactive for the players, it is likely that all the actors involved gained valuable insights into one another's' values and preferences. Many studies have attested to this (Dionnet et al. [2008]; Pahl-Wostl [2006]). The participatory mapping approach, on the other hand, allowed the information collected from a survey to be cross-checked with other stakeholders who were familiar with the context or location. For instance, during the household survey, depending on which household member was available to provide information, information may be under or overestimated. The checking allowed a level of legitimacy to be assumed. Nonetheless, during the participatory activity, males and females had different perceptions of space or spatial extent of the landscape as well as land marks in the landscapes important for delivering ecosystem services. Incorporating this information into the ecosystem service trade-offs assessment could minimize biases in recommending decisions (or solutions).

In the third case study, the exploration of the visual perception of ecosystem services through photographs (López-Santiago et al. [2014]) helped to reveal different stakeholder groups according to their values and types of knowledge. Visual surveys are quick and easy for interviewees. They provide a measure of social perception of ecosystem service delivery in a constructed metric consisting of a numeric scale associated to rating categories. Both ecosystem services with a market value and those with no market value are equally identified, thus a wide range of material and non-material links between society/individuals and nature are incorporated. Following the model of value dimensions offered by Chan et al. ([2012]), the visual method is apt to elicit market-mediated values such as those linked to provisioning services plus some cultural services such as tourism and hunting. Further, this method could also be used to elicit non-market-mediated values such as those arising from appreciating regulating services and the rest of cultural services. The concept of ecosystem services allows prioritization of self-oriented, individual, experiential and anthropocentric values, but makes less easy the elicitation of some others-oriented, holistic, existence/bequest and biocentric values.

A quantitative approach allows the objective grouping of stakeholders not only according to their socioeconomic characteristics but also regarding their links to SES, their visions of the subject under investigation, and their type of knowledge. These complement other qualitative and more participatory/deliberative tools.

Since conservation interventions require changes in human behavior to succeed (Mascia et al.[2003]), socio-cultural approaches should be considered flagship for ecosystem services assessments (Cowling et al. [2008]; Chan et al. [2012]; Martin-Lopez et al. [2007]; Oteros-Rozas et al. [2014]). Consequently, we argue that the involvement of stakeholders, as well as social scientists in research about ecosystem services, is a challenge that could be critical to target regional conservation and sustainability goals together with the well-being of human populations. A comparison of the three approaches related to depth of analysis of ecological processes, type of stakeholders that can be included, main kind of values elicited, etc. is shown in Table 1. From this study, new research questions for the future arise that we could not tackle due to time requirement and cases for comparison, such as *Can the design of a research project to use methods that allow for higher degrees of participation facilitate the elicitation of a wider variety of actors' values and perceptions? How does the degree of participation relate to the variety of perceptions and values that are revealed by participants?*

Table 1: A comparison of the three participatory methods employed in understanding ecological processes

Participatory method/case study	**Types of participationa**	**Typologies of stakeholders that can be included**	**Level and type of analysis of ecological aspects**	**Main kind of values elicited**

Participatory scenario planning in Doñana coastal wetland, Spain	Participation in information giving/ Functional participation[b]	All	Broad analysis of ecological aspects focusing more upon the interrelations among factors	Others-oriented values, group values, supporting values
Participatory mapping in rubber agroforest landscape, Indonesia	Participation by consultation/Functional participation	All	Administrative boundaries and land use types, location of water sources, production and critical conservation areas	Anthropocentric and biocentric values, group and physical values

Role playing games in rubber agroforest landscape, Indonesia	Interactive participation	Depending on the research objective/s	Abstract representation of ecological processes and dynamics	Anthropocentric and biocentric values, transformative and non-transformative values, non-market-mediated values
Evaluation of visual preferences in transhumance cultural landscapes, Spain	Participation by consultation	Depending on the research objective/s	Perception of ecosystem services delivery, some directly and some indirectly but all, linked to ecological aspects	Physical and metaphysical, biocentric, self-oriented values

[a]Based on Pretty et al. ([1995]).

[b]This method has the potential to allow the introduction of interactive participation.

Villamor *et al.*

Villamor *et al.* *Ecological Processes* 2014 **3**:22, doi: 10.1186/s13717-014-0022-9

CONCLUSIONS

We have presented three case studies that use different participatory approaches (with a strong emphasis on socio-cultural aspects) to explore how stakeholders' knowledge, preferences, perceptions, and values are integrated in social-ecological assessments. Social-ecological systems are inherently complex and include a variety of actors with different interests and values that interact. It is therefore crucial to select which participatory tools or approaches are appropriate for each specific context and objective. Moreover, recognizing and integrating the various types of knowledge and values of different actors contributes to a more comprehensive understanding of the dynamics and processes of social-ecological systems. This, we argue, is needed in the context of a search for collectively designed and supported solutions to socially rooted ecological problems.

AUTHORS' CONTRIBUTIONS

GBV, IP, and JH conceptualized and designed the paper. GBV, IP, EOR, and CLS carried out the case studies presented, performed the analysis, and wrote the manuscript. GBV, IP, EOR, CLS, and JH reviewed and commented on various versions of the manuscript. All authors read and approved the final manuscript.

ACKNOWLEDGEMENTS

The authors wish to thank all the people who collaborated on the projects presented here, especially those who participated in the research/workshops. We are grateful to the anonymous reviewers for their valuable comments.

REFERENCES

1. Akiefnawati R, Villamor GB, Zulfikar F, Budisetiawan I, Mulyoutami E, Ayat A, van Noordwijk M (2010) Stewardship agreement to reduce emissions from deforestation and degradation (REDD): Lubuk Beringin's hutan desa as the first village forest in Indonesia. Int For Rev 12(4):349-360
2. Ban NC, Mills M, Tam J, Hicks CC, Klain S, Stoeckl N, Bottrill MC, Levine J, Pressey RL, Satterfield T (2013) A social-ecological approach to conservation planning: embedding social considerations. Front Ecol Environ 11(4):194-202
3. Berkes F, Folke C (1998) Linking social and ecological systems: management practices and social mechanisms for building resilience. Cambridge University Press, Cambridge.
4. Berkes F, Colding J, Folke C (2000) Rediscovery of traditional ecological knowledge as adaptive management. Ecol Appl 10(5):1251-1262
5. Beukema H, Danielsen F, Vincent G, Hardiwinoto S, van Andel J (2007) Plant and bird diversity in rubber agroforests in the lowlands of Sumatra, Indonesia. Agrofor Syst 70:217-242
6. Bunce R, De Aranzabal I, Schmitz M, Pineda F (2006) A reviews of the role of Drove Roads (Cañadas) as ecological corridors. Wageningen, Alterra.
7. Cash DW, Clark WC, Alcock F, Dickson NM, Eckley N, Guston DH, Jäger J, Mitchell RB (2003) Knowledge systems for sustainable development. Proc Natl Acad Sci 100(14):8086-8091 http://www.pnas.org/content/100/14/8086.
8. Chan K, Satterfield T, Goldstein J (2012) Rethinking ecosystem services to better address and navigate cultural values. Ecol Econ 74:8-18

9. Cowling RM, Egoh B, Knight AT, O'Farrell PJ, Reyers B, Rouget M, Roux DJ, Welz A, Wilhelm-Rechman A (2008) An operational model for mainstreaming ecosystem services for implementation. Proc Natl Acad Sci 105(28):9483-9488
10. Daniel TC (2001) Whither scenic beauty? Visual landscape quality assessment in the 21st century. Landsc Urban Plan 54(1):267-281
11. Dionnet M, Kuper M, Hammani A, Garin P (2008) Combining rope-playing games and policy simulation exercises: an experience with Moroccan smallholder farmers. Simulation & Gaming 39:498
12. García-Martín P (2004) The history and characteristics of the Mesteña Transhumance Routes. Transhumance and Biodiversity in European Mountains' Report from the EU-FP5 project Transhumount IALE Publication Series (1):255–258
13. Gómez Sal A, Lorente I (2004) The present status and ecological consequences of transhumance in Spain. Transhumance and biodiversity in European mountains, In.
14. Gómez-Baggethun E, de Groot R, Lomas PL, Montes C (2010) The history of ecosystem services in economic theory and practice: from early notions to markets and payment schemes. Ecol Econ 69(6):1209-1218
15. Gouyon A, de Foresta H, Levang P (1993) Does 'Jungle Rubber' deserve its name? An analysis of rubber agroforesty system in Southeast Asia. Agrofor Syst 22:181-20
16. Liu J, Dietz T, Carpenter SR, Folke C, Alberti M, Redman CL, Schneider SH, Ostrom E, Pell AN, Lubchenco J, Taylor WW, Ouyang Z, Deadman P, Kratz T, Provencher W (2007) Coupled human and natural systems. AMBIO: A Journal of the Human Environment 36(8):639-649 doi:10.1579/0044-7447(2007)36[639:chans]2.0.co;2
17. Long AJ, Nair PKR (1999) Trees outside forests: agro-, community, and urban forestry. New For 17:145-174
18. López-Santiago CA, Oteros-Rozas E, Martín-López B, Plieninger T, Martín EG, González JA (2014) Using visual stimuli to explore the social perceptions of ecosystem services in cultural landscapes: the case of transhumance in Mediterranean Spain. Ecol Soc 19(2):27 http://dx.doi.org/10.5751/ES-06401-190227

19. Manzano P, Malo JE (2006) Extreme long-distance seed dispersal via sheep. Front Ecol Environ 4(5):244-248
20. Martin-Lopez B, Montes C, Benayas J (2007) The non-economic motives behind the willingness to pay for biodiversity conservation. Biol Conserv 139:67-82
21. Mascia MB, Brosius JP, Dobson TA, Forbes BC, Horowitz L, McKean MA, Turner NJ (2003) Conservation and the social sciences. Conserv Biol 17(3):649-650
22. Michon G (2005) Domesticating forests: how farmers manage forest resources. IRD, CIFOR and ICRAF, Bogor, Indonesia.
23. Montes C, Borja F, Bravo MA, Moreira JM (1998) Reconocimiento Biofísico de Espacios Naturales Protegidos. Doñana: Una Aproximación Ecosistómica. Consejería de Medio Ambiente, Junta de Andalucía, Sevilla, España.
24. Nieto-Romero M, Oteros-Rozas E, González JA, Martín-López B (2013) Exploring the knowledge landscape of ecosystem services assessments in Mediterranean agroecosystems: insights for future research. Environmental Science & Policy.
25. Norgaard R (1994) Development betrayed: the end of progress and a coevolutionary revisioning of the future, 1994. Routledge, London.
26. Oteros-Rozas E, González J, Martín-López B, López C, Montes C (2012a) Ecosystem services and social-ecological resilience in transhumance cultural landscapes: learning from the past, looking for a future. In: Plieninger T, Bieling C (eds) Resilience and the cultural landscape understanding and managing change in human-shaped environments, Cambridge University Press, Cambridge. pp 242-260
27. Oteros-Rozas E, González JA, Martín-López B, López CA, Zorrilla-Miras P, Montes C (2012) Evaluating ecosystem services in transhumance cultural landscapes: an interdisciplinary and participatory framework. GAIA: Ecological Perspectives for Science & Society 21(3):185-193
28. Oteros-Rozas E, Martín-López B, López CA, Palomo I, González JA (2013) Envisioning the future of transhumant pastoralism through participatory scenario planning: a case study in Spain. The Rangeland Journal 35(3):251-272

29. Oteros-Rozas E, Martín-López B, González JA, Plieninger T, López CA, Montes C (2014) Socio-cultural valuation of ecosystem services in a transhumance social-ecological network. Reg Environ Chang 14(4):1269-1289
30. Pahl-Wostl C (2006) The importance of social learning in restoring the multifunctionality of rivers and floodplains. Ecol Soc 11(1):10
31. Palomo I, Martín-López B, López-Santiago C, Montes C (2011) Participatory scenario planning for protected areas management under the ecosystem services framework: the doñana social-ecological system in southwestern Spain. Ecology & Society 16(1):23
32. Palomo I, Martín-López B, Zorrilla-Miras P, Del Amo DG, Montes C (2014) Deliberative mapping of ecosystem services within and around Doñana National Park (SW Spain) in relation to land use change. Reg Environ Chang 14(1):237-251
33. Pretty JN, Guijt I, Thompson J, Scoones I (1995) Participatory Learning and Action: A Trainer's Guide. IIED Participatory Methodology Series, International Institute for Environment and Development. , London
34. Rao C (1964) The use and interpretation of principal component analysis in applied research. Sankhyā. The Indian Journal of Statistics Series A :329–358
35. Reed MS (2008) Stakeholder participation for environmental management: a literature review. Biol Conserv 141(10):2417-2431
36. Reid H, Alam M, Berger R, Cannon T, Huq S, Milligan A (2009) Community-based adaptation to climate change: an overview. Participatory learning and action 60(1):11-33
37. Ruiz M, Ruiz J (1986) Ecological history of transhumance in Spain. Biol Conserv 37(1):73-86
38. Tomich TP, van Noordwijk M, Budidarsono S, Gillison A, Kusumanto T, Murdiyarso D, Stolle F, Fagi AM (2001) Agricultural intensification, deforestation, and the environment: assessing tradeoffs in Sumatra, Indonesia. Tradeoffs or synergies :221–244

39. van Noordwijk M (2011) Linking sustainability and agroforestry science to multifunctional policy action, Sixth GCHERA Conference Proceedings.
40. van Noordwijk M, Leimona B, Jindal R, Villamor GB, Vardhan M, Namirembe S, Catacutan D, Kerr J, Minang PA, Tomich TP (2012) Payments for environmental services: evolution toward efficient and fair incentives for multifunctional landscapes. Annu Rev Environ Resour 37(1):389-420 doi:10.1146/annurev-environ-042511-150526
41. Vatn A (2005) Institutions and the environment. Edward Elgar, Cheltenham.
42. Vervoort JM, Thornton PK, Kristjanson P, Förch W, Ericksen PJ, Kok K, Ingram JSI, Herrero M, Palazzo A, Helfgott AES, Wilkinson A, Havlík P, Mason-D'Croz D, Jost C (2014) Challenges to scenario-guided adaptive action on food security under climate change. Global Environ Change 28:383-394 doi:http://dx.doi.org/10.1016/j.gloenvcha.2014.03.001
43. Vihervaara P, Rönkä M, Walls M (2010) Trends in ecosystem service research: early steps and current drivers. Ambio 39(4):314-324
44. Villamor GB (2012) Flexibility of multi-agent system models for rubber agroforest landscapes and social response to emerging reward mechanisms for ecosystem services in Sumatra. Indonesia University of Bonn Press, Bonn.
45. Villamor GB, van Noordwijk M (2011) Social role-play games vs individual perceptions of conservation and PES agreements for maintaining rubber agroforests in Jambi (Sumatra), Indonesia. Ecol Soc 16(3):27
46. Villamor GB, Desrianti F, Akiefnawati R, Amaruzaman S, van Noordwijk M (2013) Gender influences decisions to change land use practices in the tropical forest margins of Jambi. Mitigation and Adaptation Strategies for Global Change, Indonesia.
47. Villamor GB, Le QB, Djanibekov U, van Noordwijk M, Vlek PLG (2014) Biodiversity in rubber agroforests, carbon emissions and rural livelihoods: multi-agent, multi-dimensional simulation tool to preview policy scenarios in lowland Sumatra. Environ Model Softw 61:151-165

48. Voinov A, Gaddis EJB (2008) Lessons for successful participatory watershed modeling: a perspective from modeling practitioners. Ecol Model 216(2):197-207
49. Young JC, Jordan A, R Searle K, Butler A, S Chapman D, Simmons P, Watt AD (2013) Does stakeholder involvement really benefit biodiversity conservation? Biol Conserv 158:359-370
50. Zorrilla-Miras P, Palomo I, Gómez-Baggethum E, Martín-López B, Lomas PL (2014) Effects of land-use change on wetland ecosystem services: a case study in the Doñana natural areas marshes, SW Spain. Landsc Urban Plan 122:160-174

Chapter 9

Assessment of Camel and Camel Milk Marketing Practices in an Emerging Peri-Urban Production System in Isiolo County, Kenya

Issack Mohamed Noor[1], Abdi Yakub Guliye[1], Muhammad Tariq[2, 3], and Bockline Omedo Bebe

[1]Department of Animal Science, Egerton University, P. O. Box 536–20115, Egerton, Kenya

[2]Animal Husbandry in the Tropics and Subtropics, University of Kassel, D-37213, Witzenhausen, Germany

[3]Department of Livestock Management, University of Agriculture Faisalabad, Faisalabad 38000, Pakistan

ABSTRACT

Pastoral camel management practices in Kenya, characterised by free herd mobility that enables efficient utilisation of rangeland resources, is progressively restricted to foraging within the vicinity of urban milk market outlets. The emerging peri-urban camel production system (PUCPS) has potential livelihood benefits to households, but adapting herd management practices responsive to market demands is a challenge with implications for its sustained development. This study assessed marketing practices in PUCPS, guided by two research questions. To what extent are (i) camel keepers involved in trading of camel stock and milk? (ii) milk hygiene practices responsive to market demands? Data was obtained from cross-sectional surveys complemented by focus group discussions. Compared to the pastoral production system, the peri-urban system exhibited greater market integration with more milk marketing opportunities, 2.4 times more (25.8% vs 62.8%) steer sales and 2.2 times more heifer purchases for breeding (12.3% vs 27.5%). Camels were sold to meet livelihood needs of the households as well as to raise cash for other direct investments. On the other hand, PUCPS growth is facing market barriers from poor milk hygiene practices. Compared to the pastoral system, the peri-urban system exhibits greater market-oriented production of camel stock and milk. Existing milk hygiene and quality practices are unlikely to meet the safety and quality requirements for urban consumers. Development of organised marketing channels and strengthening of processes that add value to milk would enable camel producers to earn more from their stock and guarantee safety and quality to urban consumers.

BACKGROUND

Camels are an important livelihood asset in the arid and semi-arid lands (ASALs) of Kenya which cover over 83% of the land mass and support about 30% (12 million) of the country's population. Camels are a source of food, cash income, transport means and have significant cultural functions to pastoral communities dominating in the ASALs (Guliye et al. 2007; MoLD 2007; Mehari et al. 2007a, b; Mahmoud 2010). These pastoral communities keep one humped camel (*Camelus dromedarius*), estimated at 2.97 million heads (KNBS 2010), and mainly concentrated in the ASALs. It is estimated that the Kenyan camel population is capable of producing between 340 and 350 million litres of milk (Faye 2007; Akweya et al. 2010a) and 10,000 tonnes of meat a year (Faye 2007). The health-promoting properties of camel milk are a strong boost for sales and, in certain regions such as the Middle East, are the drivers for intensification of camel dairying (Faye 2007).

In arid northern Kenya, camels are traditionally kept under pastoral (nomadic) production systems, characterised by low production inputs and herd/household mobility. This is a subsistence-based system utilising large mobile herds grazing on vast rangeland pasture resources. There is however a recent emergence of peri-urban camel production system (PUCPS) using milking herds grazed within proximity to urban market outlets for milk, meat and stock, where consumers and traders are members of pastoral communities (Matofari et al. 2007; Noor et al. 2012). Isiolo town in northern Kenya has a camel production system that exemplifies this emerging PUCPS, with the potential of adding value to camel milk and stock, which would improve livelihoods of the pastoral communities (LPP, LIFE Network, IUCN-WISP and FAO 2010) in the ASALs. Isiolo County has about 39,084 camels and a human population of about 143,294 (KNBS 2010).

The prevailing market-oriented camel production presents opportunities for poorer households to enhance their food and income securities (SRA 2004) in an area where annual income averages of US$217 to 301 fall below the national average of US$360 (ADF 2003). In these areas, viable alternative economic activities are lacking, resulting in a high dependency on famine relief support from government and non-governmental organisations (NGOs). Moreover, little is known about the marketing practices for camel livestock and milk in the emerging peri-urban system, which is an alternative income source. Such information would be useful for planning and implementation of targeted camel development programmes. This study was undertaken with the following objectives: (i) To estimate purchase and sale of camel stock and milk in the two systems and (ii) to determine whether milk hygiene practices in the peri-urban system are responsive to urban market requirement.

STUDY AREA

The study was conducted in Isiolo County, a typical ASAL area in northern Kenya with both peri-urban and pastoral camel production systems and a thriving camel milk trade (Figure 1). The study sites of peri-urban and pastoral areas were about 75 km from each other. Isiolo County is a semi-arid area that experiences recurring droughts with devastating losses of livestock and human lives. The rainfall pattern is bimodal but unpredictable and erratic in distribution. Long rains come in late March through May and short rains in November to December. Most parts of the County have mean annual temperatures between 24°C and 30°C (Herlocker et al. 1993). The Somali tribesmen, who speak Somali language, form the predominant camel owners in both peri-urban (80%) and pastoral (90%) systems. The Borana tribesmen, who speak Boran language, were traditionally cattle keepers, until threats posed by recurrent prolonged drought spells in ASAL areas, awakened their interest in camel keeping. They form 18.3% of peri-urban camel owners but were yet to fully adopt a pastoral camel production system, with only 10% of Borana having adopted.

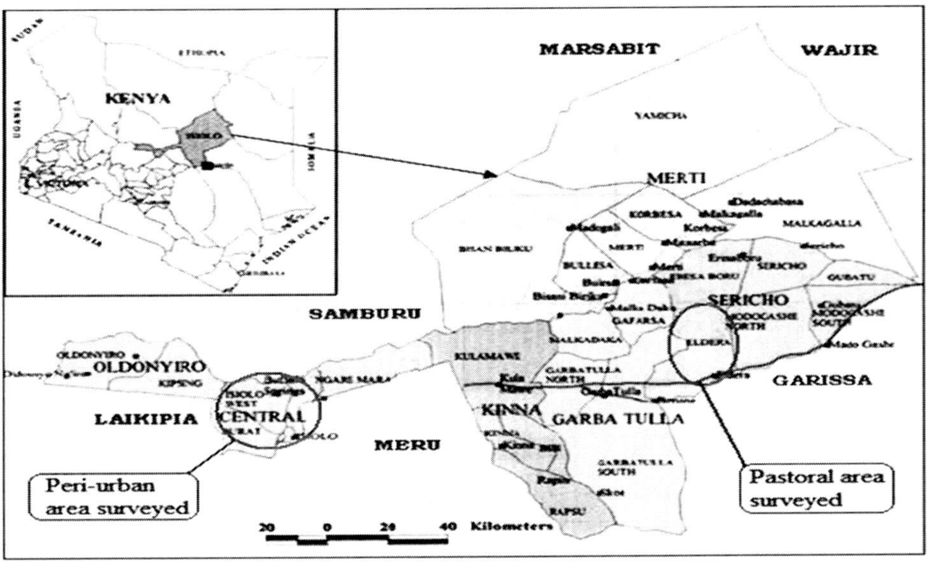

Figure 1: Map of Isiolo County (Kenya) showing the peri-urban and pastoral camel production study sites.

METHODS

Data Collection

Two cross-sectional surveys were conducted in February 2007: one within the peri-urban area of Isiolo town and the other in the pastoral rangelands of the County where the pastoral camel production system is practised and where traditional attachments to socio-cultural roles of camels remain strong. The former represented a peri-urban camel production system, where milking herds are reared within the vicinity of Isiolo town with a lot of market integration through sale of camel stock and camel milk. The latter area represents traditional camel production in a pastoral setting, where pastoralists rear their camels in the rangelands for subsistence and they move with their herds in search of pasture, water and mineral licks.

Because of high mobility, the scattered nature of pastoral camel herds and the necessity to obtain consent of the herd owners in accessing their herds, sampling in both rangelands and peri-urban sites was limited to accessible herds whose owners were willing to participate in the studies. The sampling included 70 pastoral herds and 60 peri-urban herds. The individual herds formed the sampling units, and the herd owners interviewed provided the primary data. Trained enumerators, who spoke the language of the respondents and were supervised by the authors, conducted the surveys using a questionnaire pre-tested on a small sample (10 respondents) of camel herders to test the clarity of the questions. Data collected through the surveys was a recall of the last 12 months preceding the interview period and was complemented by a focus group discussion (FGD) involving 50 participants. The participants included camel producers (22), camel milk traders (10), representatives of government departments (10), non-governmental organizations (NGOs) (2), community-based organisations (2) and local leaders (4). Camel producers and traders were the majority (70%) of the FGD participants. Specifically, data was collected on sales and purchases of camel stock and milk. Camel stocks were in four classes of heifers (before age at first calving), steers (up to 4 years of age), breeding females and breeding males. Prices and reasons for selling and buying were obtained for each case. Data on milk hygiene and marketing channels was collected during herd visits and FGD with stakeholders.

Data Analyses

Comparative descriptive and inferential statistical analyses of pastoral and peri-urban camel systems were done to test for any statistically significant difference in means of the test parameters. In this regard, camel stock purchases and sales were computed as a proportion of purchases and sales divided by the average number in the herd for each class during the year preceding the survey. The *Chi-square*test was applied to detect any significant differences in means between pastoral and peri-urban systems and Fisher's exact test used where 50% of the cells had expected counts less than 5. The proportion of animals sold and purchased was the proxy for the extent of market orientation. Frequencies for different reasons for sale declared for each animal sold within each system were calculated as a percentage of the number of responses. The average price of camel stock sale and purchase was computed for each class and subjected to a *t* test to detect significant differences between the systems.

RESULTS

Marketing of Camel Stock

The estimates of camel stock sales and purchases in the studied pastoral and peri-urban systems are presented in Table 1. Compared to pastoral producers, peri-urban producers sold 2.4 times more steer surplus stock (25.8% vs 62.8%) and bought 2.2 times more heifer breeding stock (12.3% vs 27.5%). Although the difference in the means of breeding female and male camel sales was significantly higher ($p<0.05$) in pastoral than in peri-urban system, the actual numbers involved were low. Further, there was minimal buying of the two camel classes into their respective herds; the reasons being that selling of breeding animals were usually considered suspicious in one way or another. Unless under exceptional circumstances, adult breeding females brought to market were considered as being culled for reason(s) known to the owner. Though buyers interested in meat animals have no problem so long as the body condition was good, animal breeders interested in replacement stock would have a second thought before purchasing an adult breeding female from the market.

Table 1: Sales and purchases of different camel classes by camel keepers in pastoral and peri-urban systems

Camel class	Production system	*N* (camels)	Sales (%)	Purchases (%)
Heifers	Pastoral	301	5.3	12.3
	Peri-urban	382	2.1	27.5
	Total	683	3.5	20.8
	$\chi2$ value		5.153*	23.602**
Steers (≤4 years)	Pastoral	186	25.8	2.7
	Peri-urban	204	62.8	0.7
	Total	390	45.1	1.8
	$\chi2$ value		53.610**	1.610 NS
Breeding females	Pastoral	1,143	2.6	0.3
	Peri-urban	1,217	1.5	0.5
	Total	2,360	2.0	0.4
	$\chi2$ value		3.882*	0.825 NS
Breeding males	Pastoral	156	29.5	1.9
	Peri-urban	82	17.1	0.0
	Total	238	25.2	1.3
	$\chi2$ value		4.393*	1.597 NS

*=$p<0.05$; **=$p<0.0001$; NS=$p>0.05$.

Noor *et al.*

Noor *et al. Pastoralism: Research, Policy and Practice* 2013 **3**:28 doi:10.1186/2041-7136-3-28

The stated reasons for selling different classes of camels are presented in Table 2. In both systems, camels were sold mainly for livelihood needs of the households (food, clothing, healthcare, school fees) and to raise cash for direct investments, and were rarely sold because of disease, poor performance or destocking. This scenario was reflected both in overall sales of camel stock and for the different camel classes. It was only in respect of breeding females in both systems and breeding males in the pastoral system where sales attributed to poor performance gains some prominence. The main reason for buying camels was for breeding purposes.

Table 2: Frequently mentioned reason(s) for selling different camel classes in pastoral and peri-urban systems

Camel class	**Production system**	**Total sales (*N*)**	Reasons for selling (%)				
			Livelihood needs	**Direct investments**	**Disease cases**	**Poor performance**	**Destocking**
Overall	Pastoral	140	65.0	26.4	0.7	7.1	0.7
	Peri-urban	168	67.9	22.0	1.8	6.0	2.4
Heifers	Pastoral	16	8.6	2.9	0.0	0.0	0.0
	Peri-urban	8	4.2	0.0	0.6	0.0	0.0
Steers (≤4 years)	Pastoral	48	26.4	7.9	0.0	0.0	0.0
	Peri-urban	128	60.1	16.1	0.0	0.0	0.0
Breeding females	Pastoral	30	9.3	7.1	0.7	4.3	0.0
	Peri-urban	18	0.6	3.0	1.2	6.0	0.0
Breeding males	Pastoral	46	20.7	8.6	0.0	2.9	0.7
	Peri-urban	14	3.0	3.0	0.0	0.0	2.4

Noor *et al.*

Noor *et al. Pastoralism: Research, Policy and Practice* 2013 **3**:28 doi:10.1186/2041-7136-3-28

Tables 3 and 4 present the mean prices (given in Kenya shillings and equivalent US dollars) for selling and purchasing different classes of camels in pastoral and peri-urban systems. The mean prices were significantly higher ($p<0.05$) for both sales and purchases in the pastoral system than in peri-urban system. This finding may be attributed to the fact that the quality/weight of peri-urban camels sold was lower than pastoral camels which are better fed as a result of better exploitation of wider rangeland resources.

Table 3: Mean (± SD) sale price (Ksh) of different classes of camels in pastoral and peri-urban systems

Camel class	Production system		**Mean difference**	***T* test (*p* value)**
	Pastoral	**Peri-urban**		
Heifers	17,070 ± 10,864	11,200 ± 5,575	5,870	0.282 NS
Steers (≤4 years)	11,138 ± 7,838	10,800 ± 3,082	338	0.835 NS
Breeding females	16,738 ± 10,322	13,900 ± 3,814	2,838	0.407 NS
Breeding males	20,152 ± 8,148	12,800 ± 4,532	7,352	0.044*

KSh. 69 = US$1 at the time of the study (2007). *$p < 0.05$; NS = $p > 0.05$.

Noor *et al.*

Noor *et al. Pastoralism: Research, Policy and Practice* 2013 **3**:28 doi:10.1186/2041-7136-3-28

Table 4: Mean (± SD) purchase price (Ksh) of different classes of camels in pastoral and peri-urban systems

Camel class	Production system		**Mean difference**	***T* test (*p* value)**
	Pastoral	**Peri-urban**		
Heifers	22,308 ± 16,616	13,022 ± 3,294	9,286	0.013*
Steers (≤4 years)	9,375 ± 2,428	10,000[a]	625	0.833 NS
Breeding females	15,000 ± 1,000	14,000 ± 6,164	1,000	0.797 NS

KSh. 69 = US$1 at the time of the study (2007). [a]Only three males were purchased in the peri-urban system for the same price. *$p < 0.05$; NS = $p > 0.05$.

Noor *et al.*

Noor *et al. Pastoralism: Research, Policy and Practice* 2013 **3**:28 doi:10.1186/2041-7136-3-28

Marketing of Camel Milk

Under the pastoral system, milk production was mainly for subsistence (household) consumption and calf feeding. Discussions with pastoral producers revealed that they would wish to sell milk to get money for other needs, but they do not have ease of access to markets due to distance and the poor road infrastructure. Figure 2 illustrates camel milk marketing channels from producers to consumers in Nairobi urban market. The bulk of the milk was sourced from peri-urban herds and less from pastoral herds located in areas near to reliable road network to Isiolo town (such as Kulamawe in Kinna Division and Boji in Garba Tulla Division) (Figure 2). The bulk of the camel milk leaving Isiolo town for Nairobi urban markets was fresh raw milk. Neither the producers nor the traders were value adding.

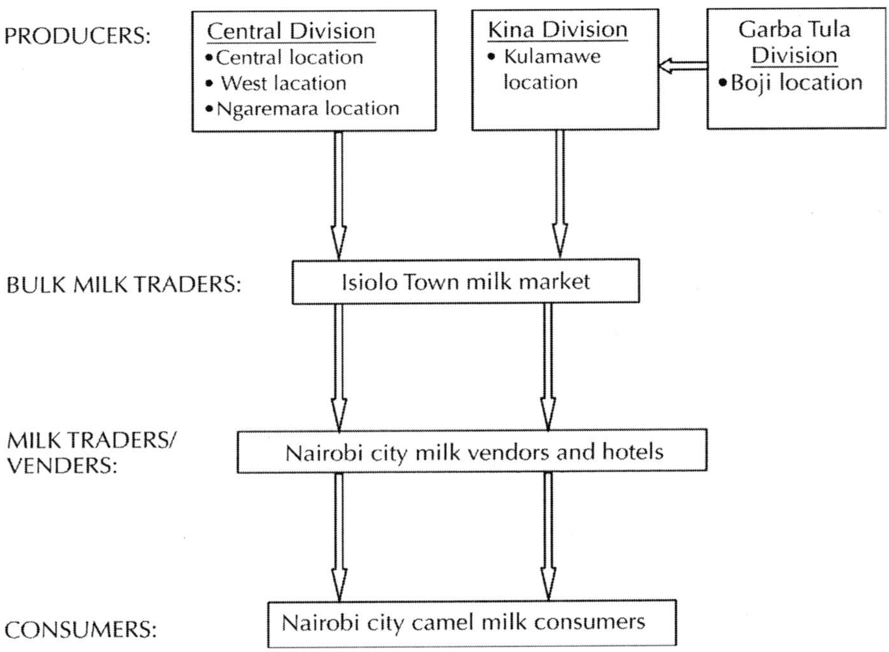

Figure 2: Camel milk marketing channels from peri-urban producers in Isiolo to consumers in Nairobi, Kenya.

Analyses of the marketing characteristics of camel milk from Isiolo peri-urban area revealed a litre of fresh raw camel milk was sold at about Ksh. 30 (equivalent to US$0.4). This was about eight times lower than the price of processed milk sold at Ksh. 240 (equivalent to US$3.5) in Nairobi urban markets. Processed camel milk sells for a lot more than unprocessed milk due to the improved hygienic quality. In Nairobi, processed camel milk is usually sold at up-market outlets to consumers who value hygienic quality and are also ready to pay more for the better quality. Milk from within the peri-urban area was transported over a distance of about 10 km to reach Isiolo town market using donkeys (94.9%), although a few producers (5.1%) with large volumes of milk use four-wheel pick-up vehicles. The milk buyers/ purchasers (100%) were milk traders (Figure 2), and most (71.2%) of them buy milk on informal contractual arrangements with producers, reflecting an organised marketing structure.

Milk Hygiene and Quality Testing

Participants in the FGD identified the main challenges in achieving clean milk production and marketing as (i) insufficient water availability during the camel milking process, (ii) milk contamination arising from saliva of the calf during suckling to initiate milk letdown prior to milking and (iii) use of plastic containers for milking, storage and transportation that are generally not easy to clean and therefore cause milk spoilage, despite washing with hot water. The use of overnight milk cooling facilities was limited to a few camel keepers who were able to transport evening milk to Isiolo town where such facilities were available. Different types of milk quality testing methods were used by producers and traders in Isiolo County. The results show most producers and milk traders in both systems (85.7% for pastoral and 100% for peri-urban system) use subjective assessments of taste and sight (colour of milk) for hygiene and quality indicators.

DISCUSSION

Marketing of Camel Stock and Milk

The study provides evidence of greater trading of camel stock and milk in peri-urban system than in pastoral system, reflecting a shift towards a market-oriented production objective. The Kenya Camel Association

(KCA 2009) reported camel prices in Kenya ranged between Ksh. 17,000 and Ksh. 35,000 (equivalent to between US$246 and US$507). The price depended on a number of factors, including age, sex, body condition and market supply and demand forces (KCA 2009). In agreement with the present findings, Mahmoud (2010) reported a vibrant and lucrative camel stock market in the northern Kenya border town of Moyale. Farah et al. (2004a) have associated observed attractive prices and incentives with pastoral household participation in the market economy. Information on camel meat consumption was not collected in this study, but camel slaughter is on daily basis in the urban centres of Kenya, indicating an increase in camel trading, and most of the slaughter was steer surplus stock and unproductive females. Sale of camel stock enables pastoral households to meet livelihood needs and raise cash for other direct investments. Similar findings were reported by Mehari et al. (2007a) for camel pastoralists in the Somali region of Ethiopia. A possible explanation for the significantly higher ($p<0.05$) mean prices (Tables 3 and 4) for both sales and purchases of camel stock in pastoral than in peri-urban system could be due to lack of adequate market information. Since pastoral producers were usually far from urban markets, coupled with the possible challenges associated with recall data, the prices quoted most probably reflect optimism.

Some authors (Chabari and Njiru 1991; Noor 1999) have reported a number of impediments to livestock marketing in the ASALs of northern Kenya including poor quality roads, lack of reliable market information, stock rustling and general insecurity, absence of consistent livestock marketing policies and hence dependency on private traders. Strengthening processes that add value to the products (LPP, LIFE Network, IUCN-WISP and FAO 2010) would be ideal interventions to enable camel producers in peri-urban systems to earn more from their stock. However, due to poor infrastructure in remote rangelands, conventional value addition processes (such as butter and cheese) are not feasible in most pastoral systems. Traditionally, camel milk is consumed either fresh or in the form of fermented milk, known as *susa* among the Somali pastoralists. It has been observed (Noor et al. 2012) that fermented camel milk is sometimes sold at almost half the price of fresh milk and therefore does not bring as much income as fresh milk. According to Farah et al. (2007), the manufacture of butter and cheese from camel milk is not a tradition in most of the pastoral societies in eastern Africa. These products were normally obtained from cow, goat and sheep milk.

The increase in commercialization of camel milk in urban niche markets observed in Kenya in the present study, was similar to trends reported in neighbouring countries like Somalia (Herren 1990; Farah et al. 2007) and Ethiopia (Seifu 2007; Mahmoud 2010) as well as in other African countries such as Djibouti, Mauritania, Morocco and Sudan (Abeiderrahmane 2013). In the PUCPS of Isiolo town, the sale of camel milk was found to be an important economic activity, attributable to the prospect of better returns arising from the increasing demand for camel milk in urban markets. The price of camel milk in Isiolo town of US$0.4 per litre was similar to the US$0.34 per litre reported by Baars (2000) for producers in neighbouring eastern Ethiopia. However, the price of milk seems not to be influenced by the distance between the production area and terminal market.

Milk Hygiene and Quality Testing

This study provides evidence that hygiene and quality practices are unlikely to meet the safety and quality requirements for urban consumers. Subjective assessment of safety and quality based on colour and taste still predominate. These are inadequate for addressing hygiene issues previously observed in camel milking and milk handling. Several authors (Farah et al. 2004b; Matofari et al. 20072013) have identified poor hygiene of the commercialised milk. Most camel milk traders buy only fresh milk from producers since most camel milk consumers prefer fresh unfermented milk. In particular, a major segment of the consumers, comprising the Somali community, believe unprocessed camel milk has medicinal properties which would otherwise be lost through heating. However, Akweya et al. (2010b) observed that camels are usually milked in poor sanitary conditions, with all the predisposing factors to diseases such as mastitis that include dust, flies and scarce water resources. In particular, lack of clean water for milking, lack of understanding of the principles of clean milk production by camel keepers and lack of overnight milk cooling facilities present challenges. Other studies (Akweya et al. 2010b; Meile2010; Wanjohi et al. 2010) have also demonstrated the presence of common milk pathogens, mostly*Staphylococcus aureus*, in camel milk. Farah et al. (2004b) also reported drinking untreated camel milk could cause gastric distress and more serious zoonotic diseases such as brucellosis, tuberculosis or typhoid. Consequently, there could be health risks associated with the consumption of raw camel milk, which could limit wider marketing opportunities, particularly amongst non-traditional camel keeping communities.

CONCLUSIONS

Compared to the pastoral camel production system, PUCPS exhibits greater market-oriented production of camel stock and milk, but this requires improved hygiene practices to guarantee safety and quality to urban consumers and achieve a better price for producers.

Given these prevailing conditions of peri-urban camel production, development agencies interested in supporting livelihoods of rural communities using the camel as a potential resource, could assist in overcoming challenges to marketing of camel milk through replacement of plastic with quality steel or aluminium containers for storage and transporting of milk, improved water supply sources within camel-keeping areas, raising awareness of camel herders about clean milking and handling practice and through establishment of overnight milk cooling facilities within production areas. Appropriate value addition to camel milk would potentially enhance household income for peri-urban camel producers. Development of organised marketing channels can facilitate adoption of hygiene practices. Development agencies need to consider supporting the establishment of milk processing plant at Isiolo to pasteurise and add value to the growing niche urban town markets in Kenya. Besides, the use of such appropriate technologies will reduce post-production losses, remove market barriers related to quality and safety standards and expand market access, thereby opening up more business and job opportunities for peri-urban camel producers.

AUTHORS' CONTRIBUTIONS

The work reported in this manuscript was a component of a wider Doctorate study for IMN. AYG and BOB were the University supervisors of the candidate, while the MT is a visiting professor from the University of Kassel, Germany and a specialist in peri-urban livestock production systems. IMN, AYG and BOB made substantial contributions to the work from conception through data collection, analysis, drafting and revising the manuscript. MT participated in drafting and revision of the manuscript. All authors read and approved the final manuscript.

ACKNOWLEDGEMENTS

The authors acknowledge the research grant for the project entitled *Enhancing Camel Productivity for Food Security and Wealth Creation for the Peri-Urban Camel Producers in Isiolo, Northern Kenya* from the Kenya Government's Commission for Higher Education. Further, we acknowledge additional funding from Egerton University, Kenya Muslims Charitable Society and field logistical support from Ewaso Ng'iro North Development Authority with headquarters in Isiolo town. Sincere appreciation also goes to the camel herders in Isiolo County for their time during interviews and focus group discussions. Permission to acknowledge has been obtained from the institutions mentioned.

REFERENCES

1. Abeiderrahmane, NJ. 2013. Camel cheese - seemed like a good idea. CreateSpace Independent Publishing Platform. http://www.amazon.co.uk/Camel-Cheese-Seemed . Accessed 13 November 2013
2. ADF (African Development Fund). 2003. Kenya ASAL-based livestock and rural livelihoods support project: Appraisal report. Nairobi: Ministry of Livestock Development archives.
3. Akweya, BA, CG Gitao, and MW Okoth. 2010a. The acceptability of camel milk and milk products from North Eastern Province in some urban areas of Kenya. Garissa: A poster presented at International Camel Symposium.
4. Akweya, BA, CG Gitao, and MW Okoth. 2010. The prevalence of common milk pathogens and antibiotic resistance of the organisms in camel milk from North Eastern Province of Kenya. Garissa: A paper presented at International Camel Symposium.

5. Baars, RMT. 2000. Costs and returns of camels, cattle and small ruminants in pastoral herds in eastern Ethiopia. Tropical Animal Health and Production 32: 113–126.
6. Chabari, FN, and GK Njiru. 1991. Livestock marketing. In Marsabit District range management handbook of Kenya. vol. II, I, ed. Schwartz HJ, Shabani S, Walther D, 111–129. Nairobi: Ministry of Livestock Development.
7. Farah, KO, DM Nyariki, RK Ngugi, IM Noor, and AY Guliye. 2004. The Somali and the camel: Ecology, management and economics. Anthropologist 6 (1): 45–55.
8. Farah, Z, S Kappeler, A Bruntse, and L Mertz. 2004b. Milk products. In Milk and meat from the camel: Handbook on products and processing, ed. Farah Z, Fischer A, 29–50. Zurich: vdf Hochschulverlag AG an der ETH Zurich.
9. Farah, Z, M Mollet, M Younan, and R Dahi. 2007. Camel dairy in Somalia: Limiting factors and development potential. Livestock Science 110: 187–191.
10. Faye, B. 2007. Focus on camels. http://www.new-ag.info/en/focus/focusItem.php?a=1273 . Accessed 13 November 2013
11. Guliye, AY, IM Noor, BO Bebe, and IS Koskey. 2007. Role of camels (*Camelus dromedarius*) in the traditional lifestyle of Somali pastoralists in northern Kenya. Outlook on Agriculture 36 (1): 29–34.
12. Herlocker, DJ, SB Shaaban, and S Wilkes. 1993. Range management handbook of Kenya. volume II, 5: Isiolo District. Nairobi: Ministry of Agriculture, Livestock Development and Marketing.
13. Herren, UJ. 1990. The commercial sale of camel milk from pastoral herds in the Mogadishu Hinterlands, Somalia. Paper no. 30. London: Pastoral Development Network ODI.
14. KCA. 2009. Kenya Camel Association (KCA) annual report. Nairobi: KCA.
15. KNBS (Kenya National Bureau of Statistics). 2010. 2009 Population and housing census. Nairobi: Ministry of Finance and Planning.
16. LPP, LIFE Network, IUCN-WISP and FAO. 2010. Adding value to livestock diversity: Marketing to promote local breeds and improve

livelihoods. Paper no. 168. Rome: FAO Animal Production and Health.

17. Mahmoud, HA. 2010. Camel marketing in the Northern Kenya/ Southern Ethiopia Borderlands. FAC Research Update 003. Brighton: Future Agricultures Consortium, University of Sussex.
18. Matofari, JW, A Shitandi, PL Shalo, NJ Nanua, and M Younan. 2007. A survey of *Salmonella enterica*contamination of camel milk in Kenya. African Journal of Microbiology Research 1 (4): 46–50.
19. Matofari, JW, PL Shalo, M Younan, NJ Nanua, A Adongo, A Qabale, and BN Misiko. 2013. Analysis of microbial quality and safety of camel (*Camelus dromedarius*) milk chain and implications in Kenya. Journal of Agricultural Extension and Rural Development 5 (3): 50–54. http://academicjournals.org/JAERD . Accessed 29 April 2013
20. Mehari, Y, Z Mekuriaw, and G Gebru. 2007a. Camel and camel product marketing in Babilie and Kebribeyah woredas of the Jijiga Zone, Somali Region, Ethiopia. Livestock Research for Rural Development 19 (49): http://lrrd.cipav.org.co/index.html . Accessed 1 May 2007
21. Mehari, Y, Z Mekuriaw, and G Gebru. 2007b. Potentials of camel production in Babilie and Kebribeyah woredas of the Jijiga Zone, Somali Region, Ethiopia. Livestock Research for Rural Development 19 (58): http://lrrd.cipav.org.co/index.html . Accessed 1 May 2007
22. Meile, L. 2010. Microbial biodiversity of camel milk and fermented camel milk products: Technology, hygiene and safety. Garissa: A paper presented at International Camel Symposium.
23. MoLD. 2007. Ministry of Livestock Development (MoLD) annual report. Nairobi: MoLD.
24. Noor, IM. 1999. A description of the pastoral camel production system in Moyale District, Kenya. : University of Nairobi.
25. Noor, IM, BO Bebe, and AY Guliye. 2012. Analysis of an emerging peri-urban camel production in Isiolo County, Northern Kenya. Journal of Camelid Science 2012 (5): 41–61.
26. Seifu, E. 2007. Handling, preservation and utilization of camel milk and camel milk products in Shinile and Jijiga Zones, eastern

Ethiopia. Livestock Research for Rural Development 19 (86): http://lrrd.cipav.org.co/index.html . Accessed 8 Dec 2009

27. SRA. 2004. Strategy for revitalizing agriculture (SRA). Ministry of Agriculture and Ministry of Livestock and Fisheries Development, Republic of Kenya, Government Printer. Nairobi: Ministry of Livestock Development archives.

28. Wanjohi, GM, CG Gitao, and LC Bebora. 2010. The hygienic quality of camel milk marketed from North Eastern Province of Kenya and how it can be improved. Garissa: A paper presented at International Camel Symposium.

Citations

CHAPTER 1

Stephanie M. Fletcher, Jodi Thiessen, Anna Gero, Michele Rumsey, Natasha Kuruppu, and Juliet Willetts, "Traditional Coping Strategies and Disaster Response: Examples from the South Pacific Region," .doi:10.1155/2013/264503

CHAPTER 2

Sanna M. Sillankorva, Hugo Oliveira, and Joana Azeredo, "Bacteriophages and Their Role in Food Safety,"International Journal of Microbiology, vol. 2012, Article ID 863945, 13 pages, 2012. doi:10.1155/2012/863945

CHAPTER 3

Daniela Toccaceli, Agricultural Districts in the Italian Regions: Looking Toward 2020, doi: 10.1186/s40100-014-0019-9.

CHAPTER 4

Adam Soliman, Duty of Stewardship and Fisheries Governance: a Proposed Framework, doi:10.1186/s40152-014-0011-8.

CHAPTER 5

Fiachra Kearney, Ryan RJ McAllister, and, Neil D MacLeod, Conservation and grazing in Australia's north-east: the bridled nailtail wallaby, doi: 10.1186/2041-7136-2-20.

CHAPTER 6

Bilal Butt and Matthew D Turner, Clarifying Competition: the Case of Wildlife and Pastoral Livestock in East Africa, doi:10.1186/2041-7136-2-9.

CHAPTER 7

Christine Rösch, Cisco Aust, and Juliane Jörissen , Envisioning the Sustainability of the Production of Short Rotation Coppice on Grassland , doi:10.1186/2192-0567-3-7.

CHAPTER 8

Grace B Villamor, Ignacio Palomo, Cesar A López Santiago, Elisa Oteros-Rozas, and Joe Hill, Assessing stakeholders' perceptions and values towards social-ecological systems using participatory methods, doi: 10.1186/s13717-014-0022-9.

CHAPTER 9

Issack Mohamed Noor, Abdi Yakub Guliye, Muhammad Tariq, and Bockline Omedo Bebe, Assessment of Camel and Camel Milk Marketing Practices in an Emerging Peri-Urban Production System in Isiolo County, Kenya, doi:10.1186/2041-7136-3-28.

Index

L

M

N

P

R

S